P9-DTI-713

Medical Tourism and Transnational Health Care

Medical Tourism and Transnational Health Care

Edited by

David Botterill
Cardiff Metropolitan University, UK

Guido Pennings
Ghent University, Belgium

and

Tomas Mainil
NHTV Breda University, The Netherlands

palgrave
macmillan

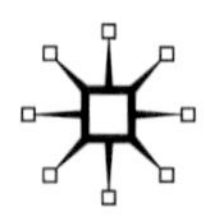

First published 2013 by
PALGRAVE MACMILLAN

Palgrave Macmillan in the UK is an imprint of Macmillan Publishers Limited, registered in England, company number 785998, of Houndmills, Basingstoke, Hampshire RG21 6XS.

Palgrave Macmillan in the US is a division of St Martin's Press LLC, 175 Fifth Avenue, New York, NY 10010.

Palgrave Macmillan is the global academic imprint of the above companies and has companies and representatives throughout the world.

Palgrave® and Macmillan® are registered trademarks in the United States, the United Kingdom, Europe and other countries.

ISBN 978–0–230–36236–9

This book is printed on paper suitable for recycling and made from fully managed and sustained forest sources. Logging, pulping and manufacturing processes are expected to conform to the environmental regulations of the country of origin.

A catalogue record for this book is available from the British Library.

A catalog record for this book is available from the Library of Congress.

Contents

Boxes, Tables and Figures

Boxes

Tables

Figures

Contributors

Krystyna Adams is a research assistant in the Medical Tourism Research Group, Simon Fraser University, Burnaby, Canada.

David Bell is Senior Lecturer in Critical Human Geography in the Department of Geography, University of Leeds, UK.

David M. Bruce is a visiting fellow in the Business School at the University of the West of England, Bristol, UK.

Rupa Chanda is Professor in Economics and Social Sciences at the Indian Institute of Management, Bangalore (IIMB), India.

Mary Choi is a research assistant in the Medical Tourism Research Group, Simon Fraser University, Burnaby, Canada.

Matt Commers is an assistant professor in European Public Health, Department of International Health, Maastricht University, Maastricht, the Netherlands.

Valorie A. Crooks is an associate professor in the Department of Geography, Simon Fraser University, Burnaby, Canada.

Keith Dinnie is a senior lecturer at the Academy of Tourism, Breda University of Applied Sciences, Breda, the Netherlands.

Henry Fraser is Professor Emeritus in the Faculty of Medical Sciences, University of the West Indies (Cave Hill).

Johanna Hanefeld is a lecturer in Global Public Health Policy at the School of Social and Political Science, University of Edinburgh, Edinburgh, UK.

Kate Hardy is a lecturer in Work and Employment Relations, Leeds University Business School, Leeds, UK.

Ruth Holliday is Professor of Gender and Culture in the Centre for Interdisciplinary Gender Studies, Faculty of Education, Social Sciences and Law, University of Leeds, Leeds, UK.

Daniel Horsfall is a lecturer in Comparative Social Policy in the Department of Social Policy and Social Work, University of York, York, UK.

Emily Hunter is a research assistant in the Institute for Interactive Media and Learning, University of Technology, Sydney, Australia.

Rory Johnston is a PhD student in the Department of Geography, Simon Fraser University, Burnaby, Canada.

Meredith Jones is a senior lecturer with the Institute for Interactive Media and Learning, at the University of Technology, Sydney, Australia.

Laura Kadowaki was most recently a research assistant in the Medical Tourism Research Group, Simon Fraser University, Burnaby, Canada.

Hannah King is a teaching fellow in the Department of Social Policy and Social Work, University of York, York, UK.

Jennifer H. Laing is a senior lecturer in the Department of Marketing and Tourism and Hospitality, La Trobe University, Bundoora, Australia.

Neil Lunt is a senior lecturer in Social Policy and Public Management in the Department of Social Policy and Social Work, University of York, York, UK.

Angie Luther is a senior lecturer in the Cardiff School of Management, Cardiff Metropolitan University, Cardiff, UK.

Melisa Martinez Álvarez is a PhD student in the Department of Global Health and Development, London School of Hygiene & Tropical Medicine, London, UK.

Herman Meulemans is a professor in the Department of Sociology, University of Antwerp, Antwerp, Belgium.

Kai Michelsen is an assistant professor in the Department of International Health, Maastricht University, Maastricht, the Netherlands.

Vincent Platenkamp is an associate professor and Director of the Centre for Cross-cultural Understanding (CCU), NHTV University of Applied Sciences, Breda, the Netherlands.

Elspeth Probyn is Professor of Gender and Cultural Studies at the University of Sydney, Australia.

Rosamond Rhodes is Professor of Medical Education and Director of Bioethics Education at Mount Sinai School of Medicine; Professor of

Philosophy at the Graduate Center, CUNY; and Professor of Bioethics and Associate Program Director of the Union-Mount Sinai Bioethics Program.

Jacqueline Sanchez Taylor is a lecturer in the Department of Sociology, University of Leicester, Leicester, UK.

Thomas D. Schiano is Medical Director, Liver Transplantation and Director of Clinical Hepatology at the Mount Sinai Medical Center and is Professor of Medicine at Mount Sinai School of Medicine, New York, NY.

Richard D. Smith is Professor of Health System Economics and Dean of the Faculty of Public Health and Policy, London School of Hygiene and Tropical Medicine, London, UK.

Jeremy Snyder is an assistant professor at the Faculty of Health Sciences, Simon Fraser University, Burnaby, Canada.

Leigh Turner is an associate professor in the Center for Bioethics, School of Public Health and College of Pharmacy, University of Minnesota, Minnesota, USA.

Wannes Van Hoof is a PhD student at the Bioethics Institute Ghent, Department of Philosophy and Moral Science, Ghent University, Belgium.

Francis van Loon is a professor in the Department of Sociology, University of Antwerp, Antwerp, Belgium.

Cornelia Voigt is Adjunct Research Fellow at Curtin University, Perth, Australia.

About the Editors

David Botterill is a freelance academic, Professor Emeritus in the Welsh Centre for Tourism Research, in Cardiff Metropolitan University, UK, and Visiting Research Fellow at the University of Westminster, UK. In 2011 he was a visiting scholar at the Cairns Institute, James Cook University, and in 2012 was appointed a visiting scientist of FAPESP, the Sao Paulo State Research Council, Campinas State University, Brazil. He has been in the vanguard of establishing Tourism Studies as a legitimate field of higher education over the past 20 years. In addition, he has worked with a number of external and industry partners, including Tourism Concern, the Wales Tourist Board, the Tourism Training Forum for Wales and the Higher Education Academy Subject Network for Hospitality, Leisure, Sport and Tourism. His more recent research has emphasised the importance of linking Tourism Studies to the philosophies, methods and concepts of the social sciences as evidenced in contributions in *Key Concepts in Tourism Research* (2012) and *Tourism and Crime: Key Themes* (2010). David is a reviewer for several publishing houses and external assessor of research quality for universities and research bodies.

Guido Pennings is Professor of Ethics and Bioethics at Ghent University, Belgium. He is also Affiliate Lecturer in the Faculty of Politics, Psychology, Sociology and International Studies at Cambridge University and Guest Professor on 'Ethics in Reproductive Medicine' at the Faculty of Medicine and Pharmaceutical Sciences the Free University Brussels. Since 2005, he is the director of the Bioethics Institute Ghent (BIG). His general interests are the ethical problems associated with medically assisted reproduction and genetics, including sex selection, reproductive tourism, gamete donation, prenatal and preimplantation genetic diagnosis. He has published approximately 180 articles in international journals and books and given numerous presentations at international congresses. He is a member of the Task Force on Ethics and Law of the European Society of Human Reproduction and Embryology (ESHRE), the Belgian National Advisory Committee for Bioethics and the Federal Commission on research on embryos in vitro.

Tomas Mainil is Lecturer and Researcher at Breda University of Applied Sciences, the Netherlands. He is responsible for the research unit 'Transnational health care in sending and receiving contexts' at the Centre for Cross-cultural Understanding and researcher at the Research Centre for Longitudinal and Life Course Studies (CELLO), University of Antwerp, Belgium. He obtained an MA in Sociology (Medical Sociology) and an MSc in Quantitative Analysis and previously worked at the University of Antwerp (Department of Sociology) and the University of Ghent (Department of General Practice and Primary Health Care) on health-related subjects. His main interests are globalisation and health, the dynamics of transnational health care and the internal and external characteristics of the transnational health user.

Acknowledgements

The editors would like to thank the management team of the Academy of Tourism, NHTV Breda University of Applied Sciences, for their generous support throughout the preparation of this volume. They would also like to thank Nick Lord for his contribution in the latter stages of the preparation of this manuscript.

1
Introduction

David Botterill, Tomas Mainil and Guido Pennings

From the perspectives of the academies associated with medicine and tourism, we will set out a position that is sceptical of the term 'Medical Tourism' yet at the same time embraces a structural bonding between two distinct and unrelated popular conceptions in the ordering of contemporary social life. Medicine and tourism have become clearly separated in contemporary popular consciousness. Medicine implies anything but a pleasurable experience, and tourism presumes a healthy disposition for participation. We argue that this popular conception of the separation of tourism and medicine ignores a historical continuity of lineage from the eighteenth-century pursuit of a 'cure' at resorts and spas to twentieth-century notions of holidays as worker welfare through global patient mobility in the quest for cutting-edge medical interventions in so-called untreatable conditions. Disciplinary divisions within the research academy have reinforced the separation between medicine and tourism in popular culture, but there is now an emergent challenge to re-think the medicine–tourism nexus. In this dynamic space and under the influence of transnational consumption, two very contrasting traditions of Western thought are now confronting each other.

International tourism markets evolved quickly following the civilian redeployment of aircraft technologies developed during the Second World War. Today those markets are characterised by a process of continual reinvention of a myriad of new segmentations in order to provide novel services to satiate consumer demand (predominantly Western but increasingly cosmopolitan). An increasingly sophisticated array of media technologies communicates the availability of custom-packaged tourist experiences. At the same time, the seemingly omnipresent desire of places to develop as tourism destinations as a driver for economic development invokes a wide range of private, public and not-for-profit stakeholders.

The medical sector has been conceived from an entirely different rationale, to care for, and hopefully cure, the ill citizen or patient. In order to do this, nations have invested heavily over several centuries in the accumulation of medical expertise and technological sophistication. Thus, the history and dynamics of tourism and medicine have been predicated on different precepts, in different time frames and under very different circumstances. However, the separation of the medicine and tourism sectors is being challenged by a new form of patient mobility – a wave of international health-care users in search of health solutions. Today, the possibilities of finding a 'treatment' elide the neatly prescribed boundaries of national health-care systems and fall headlong into the ephemeral and fleeting inventions of tourism marketeers. A new market-driven ethos in health care is forcing the previously supply-side dominated health sector to incorporate many of the demand-led industry characteristics associated with tourism. Hence, we observe a 'touristification' of health care emerging through the adoption of tourism practices but still, of course, dependent on science and technology-led medical expert treatment. The contemporary linking of medicine and tourism presents, we would argue, a dynamic window into the clash of power between global capital investment in the service industries and its impact upon notions of social welfare and national citizenship.

The movement from medical tourism towards transnational health care

Our book title suggests some ambiguity by the inclusion of two terms: medical tourism and transnational health care. In this section we explain the factors that create that ambiguity. In recent years, patient mobility across national borders has become a contextualised phenomenon. For example, European perspectives and practices are different from Asian or American ones. European nations have well-developed social security models of health care based on the common tenet of health as a social good and access to health care as a right of citizenship. In this context, cross-border health care is a more commonly used term in the debates in the European Union about satisfying the needs of patients for health care across national borders and between national health systems. More recently, some European countries, such as Germany and the United Kingdom, have recognised that health care is also ripe for market development, although such directions of policy are still clearly contained within public health goals. Policy directions such as these within European states have often been accompanied by the introduction of the term 'medical tourism' in addition to the more commonly used cross-border health care.

In a completely different context, countries in Asia and on the American continent were always more open to privatised models of the provision of health care. Therefore, on these continents, the terms medical tourism and medical travel are more commonly used, and they take on a radically different shape and form from the limited use of the term in a European context. For example, in many Asian nations the emphasis is much more on market development, and medical tourism is seen as an opportunity to drive economic prosperity in the region. Thus, medical tourism in Asia is characterised by governments as a growth industry receiving increasing numbers of international medical tourists. The markets of medical tourism are also increasingly intra-regional, with patients travelling within regions, for example, from Australia to Thailand or, in Latin America, from Columbia to Brazil.

Medical tourism is thus more associated with the provision of health care on the basis of out-of-pocket fee-for-service which, in turn, locates medical tourism as an invisible export, close to other forms of tourism, and with medical patients treated as tourist consumers. However, as we show in the chapters of this book, the distinction between the old terms – medical tourism and cross-border health care – is now less easy to draw. We propose therefore a new overarching term in this book's title – transnational health care. The use of this term would aid both the consumer of medical tourism and the citizen of health-care systems to more easily recognise the emerging set of transnational structures and networks that seek to serve all patients. Further, we propose that transnational health care could become a mature global patient mobility framework, which builds on a logic of transnational health regions (regional development as a vehicle for patient mobility), transnational organisations (such as hospital chains and insurance schemes) and sustainable health destination management (governmental steering of the development of patient mobility).

Tourism studies academics and industry practitioners tend to argue in terms of market development and treat medical tourism as a tourist activity with products, services and customers, to the exclusion of the wider implications of medical tourism for national health systems. Professionals within the health-care sector start from very different positions compared to their tourism counterparts. Medical practitioners, for example, debate about medical tourism and cross-border health care in terms of medical quality and the ethical consequences for international patients within health-care systems. Whatever the differences between these two communities of practices are, we suggest they would do well to begin to adopt a more inclusive way of seeing one another's interests. Thinking in terms of transnational regional networks might make collaboration possible between a full range of stakeholders who would gain from better serving local and global patients: as both citizens and as consumers. As a contribution to these arguments we invite the reader to think about each chapter from a triad of questions:

1. What are the public health perspectives?
2. How is market development represented?
3. Is there any evidence of integration of these positions?

The critical reader will observe that, at this stage in its formation, evidence of the integration implied by our model of transnational health care is difficult to find.

The main divide in the field is that between 'medicine' on the one hand and 'tourism' on the other. The ultimate purpose of this book is to bridge the divide. When starting on this book, the editors, as representatives of the two sides, found out that they knew almost nothing about each other's perspective on medical tourism. The main benefit resulting from the bringing together of different disciplines as presented in this book may lie in the fact that perspectives, and the underlying assumptions, are brought to the fore and confronted. The perspective adopted to present the phenomenon of medical tourism cannot be neutral or objective. The goal is not to select 'the' right or most appropriate approach. There is no 'view from nowhere': every researcher inevitably has to adopt a position from which to study medical tourism. However, being aware of one's own and the others' perspectives, and about the implications and restrictions of the perspectives, is already a great advantage. At the same time, it is an uneasy mix because of the ongoing discussions on how a 'medical tourist' should be defined. This is related to the question of 'to what extent a "touristic" purpose or element should be present'. Those who see the cross-border patient as just a patient looking for treatment abroad will object to the term 'medical tourist'. As soon as this patient chooses his destination, in part because of its attractiveness as a vacation resort, the term is perfectly sensible. The same ambiguity exists on the supply side: some countries may try to attract foreign patients by playing the tourism card (the non-treatment related aspects) while others may just attempt to pull in as many patients as possible. In the latter case, it does not matter where the clinic is located. Still, even in the latter case, tourism strategies and knowledge may be used to obtain results.

The basic element underlying medical tourism is unavailability: people need or desire something that is not on offer at home or that is not on offer in the right conditions. These conditions include, amongst others, availability within a reasonable time, affordability and good quality. The 'something' that people want may be services (high-tech or standard), body material (organs, eggs, etc.) or specific interventions. This basic structure can be presented by several dualities: supply and demand, home country and destination country, patient and provider, and so on. These perspectives are characterised by different focuses in terms of patient autonomy, state obligations, commercialisation, globalisation, the nature of medicine and health, and so on. Fundamental choices will have to be made. Will we try to stop the commercialisation of health care and put a maximum effort into maintaining equitable access to health care at home? Will we give in to the pressure, largely originating in budgetary limitations and ever-growing medical needs and possibilities, by simplifying and facilitating cross-border movements while controlling and monitoring the evolution to prevent derailments?

While the fundamental questions are largely based on ethical values, evidence will be needed on whatever decision is made in order to make sure that the actions performed, the organisational structures put in place, really express the choices. Middle ways that compromise between the market and the public health-care system, and between consumers and patients may be found. The choice about where to place the balance is ours to make.

Structure of the book

The structure of the book organises the contributions of authors into three parts. The first two parts reflect the established divide between tourism and medicine referred to above. Part I – 'Tourists as Patients' – foregrounds the shift in identity from the tourist to the patient, largely from the perspective of the conceptual models and practices of tourism consumption. In Part II – 'Patients as Tourists' – the semantic shift in the opposite direction displays the very different preoccupations of health researchers. In Part III – 'Entanglements with Medical Tourism: Policy, Management and Business Responses' – we use the word entanglements to consider how the transition from medical tourism to transnational health care takes its form within specific policy, management and business issues.

Part I – Tourists as Patients

In Chapter 2, David M. Bruce reminds us that travelling to 'take the waters' was a preferred option over and above medical interventions closer to home because, prior to around 1850, such intervention was rather more likely to kill than cure. Through the pages of guide books used by tourists as patients, Bruce paints a vivid picture of aspects of health, sickness and death and of tourism in the nineteenth century. In Chapter 3, Cornelia Voigt and Jennifer H. Laing bring the historical continuity of tourism and medicine up to date with a twenty-first century conceptual mapping of the differences and overlaps between wellness and medical tourism providers. They provide background information on the broader environment of health care and Western consumer culture that have influenced medical and wellness tourism providers. They present five provider models where the lines between medical and wellness tourism have been blurred, and they discuss areas for future research.

In Chapter 4, Jeremy Snyder and his team of co-authors provide a study of non-resident health care on the popular tourist destination island of Barbados. Barbados has an established history of providing health care to ill vacationers, other Caribbean residents and, more recently, the small but growing numbers of medical tourists seeking fertility care. Drawing upon its experience in developing tourism services, the Bajan government is aggressively promoting the development of a medical tourism industry, including the facilitation of the development of a mid-sized private hospital catering primarily to medical tourists. The authors conclude that the development of medical tourism services in Barbados carries with it not only the potential for some economic gains but also significant risks to health equity if mismanaged or left unregulated. The final chapter (Chapter 5 by Angie Luther) in this part of our book examines tourists with serious disabilities. Luther provides a sensitive and illuminating account of how the medical condition of cervical spinal cord injury impacts the tourism aspirations of people with a serious disability. Through the lived experiences and perceptions of individuals with cervical spinal cord injury, Luther provides insights into the risks taken by tourists who are patients in their everyday lives. This chapter closes by highlighting the possible policy implications that could facilitate and/or enhance participation in tourism for individuals with cervical spinal cord injury and other similarly disabled groups.

Part II – Patients as Tourists

Cosmetic surgery tourism is one of the most popular instances of medical tourism. In Chapter 6, Ruth Holliday and co-authors explore the connection between the mental renewal offered by a touristic trip and the physical renewal that follows a cosmetic intervention. They also emphasise the tension between the anticipated feeling of pleasure linked to the touristic place and the anticipated feeling of pain linked to the operation. Their analysis shows that clinics use images of the country and culture to connect with elements that attract patients, such as quality of care, low cost, hospitality and cultural proximity. More than for other forms of medical travel, this type of cross-border treatment is built on myths and stories about particular locations.

Reproductive tourism is a peculiar form of medical travelling as it is frequently the result of national legislation. Certain applications of medically assisted reproduction and genetics are forbidden in some countries and allowed in others. This legal diversity creates several problems and opportunities. In Chapter 7, Wannes Van Hoof and Guido Pennings focus on two specific controversies: the problem of payment and donor anonymity within the practice of gamete donation, and the problem of exploitation and comity (incompatibility of legislations in different countries) within the context of international commercial surrogacy. These problems are the result of divergence in ethical evaluation and are hard to solve. The evolution of cross-border reproductive care is difficult to predict as it depends on several dynamic conditions such as scientific developments, ethical and religious convictions, and changes in legislation.

The most dramatic form of medical travelling takes place in the context of organ transplantation. People desperate for an organ and confronted with a long waiting list at home seek a solution abroad. These decisions are contentious mainly because of the ethical context in which donor organs are obtained in many developing countries. Both the exploitation of poor people and the use of executed prisoners as organ donors in the People's Republic of China violate basic ethical principles. In Chapter 8, Thomas D. Schiano and Rosamond Rhodes analyse the numerous statements issued by professional societies on transplant tourism. They also scrutinise the arguments in the debate on payment for organs. They conclude that the first task of the societies is to adopt measures to guarantee the health and safety of the donors.

In Chapter 9 Tomas Mainil and his co-authors point out that within the European context the patient is as much a citizen as he is a consumer. The recent Patients' Rights Directive expresses this idea. Based on this duality in the patient, a typology that incorporates and covers various other typologies suggested in the field is proposed. This important aspect of the transnational patient has implications for the policy discussion on the effects of transnational health care for the health-care systems. Other considerations, such as the willingness of people to seek health services abroad, socio-economic differences between countries and the general economic climate, also play a role. Finally, a number of scenarios regarding the future evolution of cross-border medical treatment are presented.

Part III – Entanglements with Medical Tourism: Policy, Management and Business Responses

The final part of this book ends with the contributions of scholars from different disciplines who are tangling with medical tourism. Contributions are made from disciplines such as ethics, philosophy, medical sociology, health economics and social policy. Taken together, the chapters in Part III attempt to map the full complexity and challenges of medical tourism.

Chapter 10, by Leigh Turner, presents a comprehensive overview of the medical tourism industry in Canada. While doing so, he distinguishes different types of companies playing in the field, such as medical tourism companies, cross-border medical travel facilitators and private health insurance companies. A detailed analysis is provided of the destination countries, the health services marketed, the marketing message and the additional services that are offered. The medical tourism industry grows despite a publicly funded health-care system that gives access to medically necessary treatment. Shortcomings of the system in combination with 'experimental' procedures not offered within the system can explain why people seek care outside the country.

Guido Pennings begins Chapter 11 with the assertion that access to health care is a human right. As a consequence, governments have the task to guarantee this right. He argues that medical tourism should be seen as a health-care system reform and should be evaluated by means of Daniels' benchmarks of fairness. However, the ethical evaluation is rendered highly complex because the results differ depending on the ethical theory that one adopts. Generally speaking, four theories on the just distribution of scarce resources can be distinguished: utilitarianism, egalitarianism, prioritarianism and sufficientarianism. The main discussion focuses on the prioritarianist position since this is the main concern expressed in the literature. Medical tourism should not make those who are already worst off (poor patients in developing countries without access to basic health care) even worse off. The brain drain resulting from the inflow of foreign patients is taken as an illustration of this approach. He finally concludes that governments should regulate and control the developments by adopting measures to guarantee fair access to good-quality health care.

In Chapter 12, Tomas Mainil, Vincent Platenkamp and Herman Meulemans link the legacy of Jürgen Habermas' ideas with transnational health care, and in so doing make the case that cross-cultural management in medical tourism should be a professionalising exercise. In his later work, Jürgen Habermas drew a distinction between communicative and strategic action. The authors demonstrate this distinction within medical tourism by juxtaposing historic and dialogical ways of communicative action – the 'lifeworld' of patient and medical doctor interactions – with the strategic for-profit ways of action in the medical tourism marketplace. They argue that by developing a deeper cultural understanding and sensitivity among hospital staff who work with international patients, a better balance between Habermas' two lines of action could be achieved. Furthermore, such management intervention could contribute to medical tourism becoming a more sustainable practice.

In Chapter 13, Melisa Martínez Álvarez, Richard D. Smith and Rupa Chanda take the entanglements into the developing world. These authors show that low- and middle-income countries providing medical tourism services may indeed benefit from generating foreign exchange, attracting – and retaining – health professionals, and improving facilities and quality of care. But these countries also risk diverting scarce resources to cater for foreign patients who can bring in higher revenues, thereby neglecting the needs of the local population. The authors analyse three types of trade agreements that countries can engage in when providing medical services to international patients: multi-lateral, regional and bi-lateral. Bi-lateral trade offers countries the greatest scope to capitalise on the benefits and reduce the risks of engaging in medical tourism, as seen in a case study from a potential UK–India relationship.

Daniel Horsfall and his author colleagues focus on the Internet as a source of data for understanding medical tourism. In Chapter 14 they research dental tourism websites that target potential medical tourists. Their analysis shows that commercial sites aimed at people seeking dental treatment abroad generally appear extremely professional. This apparent professionalism of dental tourism sites masks the fact that important information is often missing from sites. Consequently, dental tourism consumers are unlikely to be fully informed of all aspects of the dental tourism process.

Finally, the range of features adopted by dental tourism sites to engender trust is broadly couched but often, in effect, meaningless. The authors make it clear that addressing the poor quality of information on dental tourism sites is extremely important to assure quality of treatment; however, for many reasons a regulatory approach is neither viable nor even desirable.

In the final chapter of Part III, Chapter 15, Tomas Mainil and his colleagues introduce the idea of a destination management framework for transnational health care. The chapter begins with a consideration of the definitions and concepts that inform an analysis of transnational health care, governance and sustainability. A model is constructed drawing upon the building blocks of destination management, specifically stakeholder, ethical and branding theories, to demonstrate how the linkages between destination management and transnational health care can be constructed. Case study examples demonstrate how regional development in relation to health and health care is an active practice in the European Union.

Part I

Tourists as Patients

2

Sickness, Health, Tourism and the Ever-Present Threat of Death: Nineteenth-Century Spa and Seasonal Travel

David M. Bruce

In this chapter I will provide information on the following:

- 'taking the waters' before and into the nineteenth century;
- aspects of nineteenth-century health, sickness and death;
- aspects of nineteenth-century tourism;
- guidebooks used by patients in the nineteenth century:
    - Mariana Starke,
    - John Murray,
    - Karl Baedeker;
- the Case of Bagni di Lucca (The Baths of Lucca); and
- nineteenth-century valetudinarian and invalid travellers.

Introduction

If scientific medicine can take as its seventeenth-century starting point the discovery of blood circulation by Harvey and Hooke's early (self-)experimentation with various substances, then arguably it took about two and a half centuries or eight generations for doctors to become *less* rather than *more* likely to kill their patients. The anxiety engendered by medics gave full scope for the more holistic type of cure represented by the spa to complement and even surpass the experimental Western science of medicine. 'Taking the waters' was a good alternative to the gamble of medical treatment and in addition spa towns also attracted doctors (as well as quacks).

By looking at places, patients and potential physicians, and the networks they inhabited, this chapter suggests a figurational-type approach (Dunning,

1996) to drawing a picture of nineteenth-century medical practice associated with travel. It draws on an extensive (but only recently published) journal from the middle of the century. The context for that journal will include how the guidebooks were used by 'valetudinarian' (as well as other) tourists of the time before the Great War. A 'valetudinarian' was defined as 'a person in weak health, esp. one who is constantly concerned with his own ailments; an invalid'. The 1787 example given in the Shorter Oxford English Dictionary says it all: 'Everyone knows how hard... it is to cure a v' (1973, p. 2448).

The literature basis for this chapter combines the histories of leisure and tourism (Towner, 1995; Koshar, 2002; Mullen and Monson, 2009) with some allusions to recent material from tourism studies (Butler and Russell, 2010; Urry and Larson, 2011) and reference to a standard and recent history of medicine (Magner, 2007). Contemporary guidebooks (Starke, Murray and Baedeker) will be referred to without intermediaries but with some reference to Parsons' general work *Worth the Detour* (2007); biographies of nineteenth-century characters also feature as do quotations from contemporary reports on the subject, including the *London Illustrated News, a Parliamentary Report* and Encyclopaedias – *Britannica* (1911) and *The Grocer's* (1911).

The chapter's six sections reflect the bullet points of the coverage and include a case study of a spa resort in Italy (Bagni di Lucca); the chapter concludes with some general comments. The approach of textual analysis and quotation is informed by and may be akin to that of the figurationists (Dunning, 1996), but the intent is introductory and to suggest areas of further research rather than claiming to be comprehensive or definitive.

'Taking the waters' before the nineteenth century

Although there was a great expansion of spas and their use by long-distance visitors in the nineteenth century, their existence had been known and they were frequented centuries before. Their antecedents are sketched out back to the time of the Roman Empire and after the Renaissance links were identified to the grand tour, when aristocratic North Europeans discovered Italy and Western Europe on educational 'gap years' (Trease, 1967).

The origin of most mineral waters, filtered through soils or 'geezered' up from deep in the Earth's crust, means most are in areas of current or past volcanic activity. The 1911 Encyclopaedia Britannica discusses their origin and worldwide coverage (Unsigned, 1911). Just about the earliest recognisable guidebook is by Pausanias, the Greek of the second-century AD, a follower of Aesculapius and so probably a Doctor. Written and circulated for wealthy and Greek-reading Romans, it showed the way to and described the godly shrines and sacred springs of ancient Greece (Levi, 1971; Bruce, 2010). Ever since there has been a tradition of visiting hot and mineral water springs in Europe.

At Baiae near Naples the weird volcanic Phlegraean Fields had long been known to be associated with hot sulphurous springs and are finely depicted by Georg Hoefnagel in the sixteenth-century Atlas of Braun and Hogenberg (Fuessel, 2008, p. 268–9). Originally associated with the gods, knowledge of the healing and soothing properties of water treatment never fully died out at the various *Thermae* (hot water spas) of Italy and even as far afield as, for example, Aquae Sulis – the waters of Sulis (Minerva) at Bath – but largely fell into disuse outside the Empire itself. After the fifth century basically that meant the eastern or Byzantine Empire, eventually to become the Ottoman Empire with its *hamans*. So 'Roman' became renamed 'Turkish Baths' and were brought back to the awareness of Central and Western Europe after Hungary was re-conquered by the Austrian Emperors (who were also Kings of Hungary) after the 1680s: Budapest remained and remains to this day a centre for balno-tourism.

The Grand Tour of the Eighteenth Century had a by-product in reviving the interest of North Europeans in spa use and indeed the South Netherlands (modern Belgium) resort of Spa gave its name to all such watery resorts. Its near neighbour Aachen demonstrated, under its French name Aix-la-Chapelle (usually used in English until the twentieth century), how for many spas health was far from being the only or even principal source of attraction for the visitor.

Nineteenth-century guidebooks naturally combined both culture and cure but were more reticent (after the early Murrays), often to the point of total omission of the subject of pleasure. Ephemeral and only rarely surviving are the 'Babylon' guides to prostitutes and other city vices (Perrottet, 2012), but such are quoted as the source of a revealing if less than fully verifiable source on a western Prussian spa town:

> Aachen became attractive as a spa by the middle of the 17th century, not so much because of the effects of the hot springs on the health of its visitors but because Aachen was then – and remained well into the 19th century – a place of high-level prostitution in Europe. Traces of this hidden agenda of the city's history are found in the 18th century guidebooks to Aachen as well as to the other spas; the main indication for visiting patients, ironically, was syphilis; only by the end of the 19th century had rheuma become the most important object of cures at Aachen.
>
> (Wikipedia, 2012)

That Aix-la-Chapelle (as Aachen was known in English as well as French) should therefore be the location for three major pan European congresses and treaties (1668, 1748 and 1818) is explicable but just why, may we enquire, did Wilkie Collins think it and, according to his biographer, find it effective for grossly (perhaps gout) swollen feet, in 1863 (Ackroyd, 2012).

Murray's confirmatory comments (1840) on German watering-places are quoted below.

Spas with their differing water constituents and temperatures might have been seriously matched to curing different ailments. When sea resorts (thalassotherapy – see Parr, 2011) and summer mountain retreats joined, or cashed in (Swiss or Italian summer retreats such as Bagni di Lucca), it is clear that holiday or pleasure motives as much as medical ones predominated before the nineteenth century. Bath in England was one of the most well-developed (and now researched) eighteenth-century spa towns. Towner (2010) describes and analyses it well in his recent book chapter on Beau Nash, an early 'Giant of Tourism'. 'However a restful, restrained stay [for a cure] ... would draw Nash's scorn' (p. 28) – his wealth came from gambling.

Medical and morbidity context

'Eighteenth Century physicians were unlikely to meet a challenge like yellow fever [for example] with such timid methods as rest and fluids' (Magner, 2007, p. 312). Magner argues the historian of medicine when a famous doctor in the 1790s did worse than nothing when faced with an epidemic in Philadelphia. The chapter in Keats' biography that ends with him seeing that 'that drop of blood is my death Warrant. I must die' (Gittings, 1968, p. 380) is instead followed by him preparing to and setting off to Italy – described as a standard practice for a (rich) consumptive of 1818. What else could he realistically do?

The fragility of life at the time meant that death was ever present – at childbirth, in childhood, from little understood epidemic diseases, from the living conditions, smoke and filth of old and industrialising cities as well as from frequent accidents. The anxiety of patients resorting to the medical profession was often well founded. Before Simpson's discovery and development of chloroform as an effective anaesthetic, surgeons were popularly known as sawbones or (in America) 'armed savages' (Magner, 2007, p. 481). Doctors often favoured aggressive intervention like bloodletting over proper nursing care. Even Simpson's chloroform was not initially always to the benefit of patients (though much valued in childbirth by, for instance, Queen Victoria). With anaesthetics, surgeons were tempted to operate more freely but often went on with earlier time-saving short cuts such as keeping scalpels in pockets and held even knives conveniently in their mouth mid-operation. Florence Nightingale and her wise obsession with orderly nursing and cleanliness only began with the Crimean War (1854) and it took time to become generally effective in hospitals. Still ignorant of the existence of germs, ignoring basic cleanliness, far less hygiene, not until Lister's demonstration and wide promulgation of the use of antiseptics would those major killers, collectively known as 'hospitalism' ('a unique nineteenth-century plague' (Magner, 2007, p. 484)), be addressed and conquered.

Meanwhile, the *Illustrated London News* (1847 but whole century) was reporting almost weekly fatal rail accidents and even a formal UK Parliamentary return (Bowring, 1860) recorded 25 passenger train accidents within 6 months – killing 12 and injuring 143 (including rail staff), whose injuries as like as not killed them later from infections contracted during treatment but the death itself went unreported.

Round about 1892 then, when Lister retired with a peerage, is probably a fair date for recognising when Western medicine began to save more patients than it killed. It was also around the time when the agents for malaria and yellow fever were isolated and linked to the mosquito. Until at least that time, with medicine and surgery so obviously dangerous and transport risks so casually taken on, travel to reduce the well-perceived risks of climate and foul air was a natural option.

Aspects of nineteenth-century tourism

The 'educational' grand tour of the late eighteenth century was for the pleasure seeking, even if family despatched, gap-year man, but the seeker of culture in Italy also sought out a climate to favour recovery or amelioration of fell diseases. Tourism in the nineteenth century was de-differentiated within itself, even if not post-modern in Urry's terms (Urry and Larsen, 2011). Laurence Sterne in *A Sentimental Journey* might have tried to find three

> general causes – Infirmity of body, Imbecility of mind, or Inevitable necessity. The two first include all those who travel by land or by water, labouring with pride, curiosity, vanity, or spleen, subdivided and combined *in infinitum*. The third class includes the whole army of peregrine martyrs; more especially with the benefit of clergy . . . as delinquents . . . – or young gentlemen transported by the cruelty of parents and guardians, and travelling under the direction of governors recommended by Oxford, Aberdeen, and Glasgow.
>
> (Sterne, 1764/1911, p. 13)

With travel still slow and transport still relatively more expensive than accommodation, nineteenth-century sojourns abroad were generally long but there was substantial even revolutionary expansion associated with the democratisation of tourism from an aristocratic 'education' by travel on the *Grand Tour* to bourgeois commercial consumption. Post Roads and Military Roads were, by the end of Napoleon's Empire, a network across Europe unmatched since Rome; steam – initially ships and river boats and then of course the railways – made further radical changes and introduced reliability of travel for the first time (Bruce, 2010).

When Ruskin travelled with his parents in pursuit of culture, he was harbouring dark thoughts about Murray's handbooks and the rail-bound tourism that was fed by them. Nineteenth-century cultural tourism was spawned by this often antagonistic intercourse of 'high' culture and cheaper more reliable travel – high culture was moving from the aristocratic to the bourgeois; tourism from the grand tour to the Cooks package though Thomas Cook himself had little effect on travel for health (Walton, 2010).

Cultural tourism arguably 'constructed' by, certainly developed as a special form of the 'gaze', John Ruskin (Hanley and Walton, 2010) was also the companion to invalid travel and the companion's relief. Much as Ruskin disliked them, these were the times of the steamship, of the early railways and the spreading industrial revolution. Throughout his multi-journey and often valetudinarian life, he remained in denial about his own tourism, an anti-tourist in Buzard's terms (1993); he saw himself as the cultured aesthete, who just happened to be obsessed with Italian and classical culture and therefore constrained to visit Italy repeatedly. His art historical works were summarised, even by himself, to become museum 'vade mecums'. But he first went to Italy with his mother and family for his delicate health. Either it worked or his health, even if not his mental or psychological wellness, improved and gave him 81 years until 1900. Certainly he continued to visit English spas, seaside resorts and Switzerland for health as well as his love of nature; for art he visited and wrote about particularly Venice and Florence. Ruskin's 'high' culture, mediated by Murray and Baedeker guides defined a 'Golden Age' of cultural tourism, which ended with the guns on the Danube in August 1914 (Bruce, 2012).

Starke, Murray and Baedeker: Guidebooks used by patients in the nineteenth century

Murray and Baedeker guidebooks directed the course of tourism for their independent traveller users from the late 1830s onwards, but before them Mariana Starke, writing her guides in the early years of the century (up to 1838), paved the way. Because she is less well known and arguably more important than either of her successors, especially for valetudinarian tourism, this chapter gives her a special focus.

Mariana Starke

Mariana Starke was the mother or nanny of all who followed in the English-speaking world and influenced the German and French guidebooks. Originally a modestly successful playwright, she wrote guidebooks in the years after 1815 based on her earlier (1800) *Letters from Italy*, which was translated into German and later editions were published by John Murray II. John Murray III used the format as the basis of his guides from 1836 onwards, and Karl Baedeker in his early collaboration with Murray picked up much of the same approach.

Starke was born in 1762 and taken abroad in 1792 with her invalid father who died of consumption in 1794. As her mother's companion roaming Italy for six years to 1798, Starke witnessed Napoleon's invasion of Italy's cultural cities – Nice (then Savoyard), Genoa, Pisa, Florence and Rome. Retreating before the French advance – Britain was at war with France – she remained royalist and anti-Jacobin throughout. Her continual (after the end of the Napoleonic Wars) travel to and around Italy spanning a further 25 years until the 1830s confirmed her authority (DNB, 1909).

Letters from Italy (Starke, 1800) were specifically collected and published for the assistance of English (or British) health travellers. As the original title page puts it, after puffing her reports of revolutions and mention of her descriptions of the arts 'WITH INSTRUCTIONS *for the use of Invalids and families*, who may not choose to incur the Expence [sic] attendant upon travelling with a COURIER'. Her extensive reports of the revolutions – always from the viewpoint of the about–to-be-destroyed *ancient regimes* – are little read now, but those 'INSTRUCTIONS' including overwhelmingly detailed accommodation and pricing information were developed and updated for her guidebooks.

As an indication, in the contents for Chapter 1 the Appendix to her penultimate edition (Starke, 1837, p. 467) is quoted in full, together with excerpts from the Appendix itself.

CLIMATES – PASSPORTS, Etc

Climates

- Invalids cautioned against exposing themselves to the influence of the Sun
- Newly built houses, and houses not built on Arches, unwholesome
- Ground floors healthy only in summer
- Best Winter Situation for Invalids
- Eligible Situations during other Seasons of the Year
- Naples and Lisbon liable to destructive Vicissitudes of Weather
- Barcelona, Valentia, and Alicant recommended during Winter

Passports

- Other requisites for Travellers, on leaving England
- Means of preserving Health during a long Journey
- Bargains with Innkeepers, etc.

'CLIMATES OF THE CONTINENT' are covered in a thousand words, concentrating on Italy and especially recommending 'A more temperate summer climate [than Florence or Pisa], which] may be found at the Baths of Lucca, where the thermometer rarely rises above 78o' (p. 468, column 1). PASSPORTS require a further 400 words – very complicated in the 1830s with the need for multiple counter-signatures from British ambassadors as well as foreign diplomats 'to avoid the trouble and detention... which frequently occur at Paris' (p. 468, column 2).

A further 1500 words itemise 'OTHER REQUISITES FOR TRAVELLERS', including everything from 'an active intelligent English Man-servant, who understands how to grease and chain wheels, and likewise to load and take care of English carriages' to 'Bramah locks for writing desks and coach seats' and 'Shuttleworth's drop-measure, an article of great importance; as the practice of administering active fluids is dangerously inaccurate...James's powder – salvolatile – sulphuric acid – pure opium – liquid laudanum – ipecacuanha – emetic tartar' and so on.

Self-medication was clearly an important part of travel; opium was not yet outlawed. Perhaps for the invalid's companion (herself), she extended well beyond advice on spas. Meticulous in her detail, she became indispensible and famous, even notorious, as the English traveller's companion, guide and friend. Dependency could become risible in the case of an 'English Historian' lampooned in the pages of Stendhal's 1839 *Chartreuse de Parme*.

> In Parma...to write a history of the Middle Ages....he refused to pay for the merest bagatelle without looking up the price in the travel guide of a Mrs Starke, which has gone into a twentieth edition because it lists for the prudent Englishman the price of a guinea fowl, an apple, a glass of milk, etc., etc.
>
> (Stendahl, 2006, p. 236)

Possibly it was Ruskin or Macaulay whom Stendahl ridiculed: Ruskin was a medievalist; Macaulay's own journal puts him near that part of Italy (researching his *Lays of Ancient Rome*) when Stendahl was thinking about the novel, and Macaulay's journal shows his use of Mariana Starke. Macaulay meanwhile was reporting on how the Italians who feared and despised her made cartoons of the absurd English using her guide – 'I am amused today by seeing in a shop [in Rome] a set of caricatures reflecting on the way in which the English appeal to Mrs Marianne [sic] Starke on every occasion: 'Voi volete ingannarmi – Madama Starke he stabito il prezzo del vino d'Orvieto tre paoli' ('You want to cheat me – Madame Starke has fixed the price of Orvieto wine at three paoli')' (Macaulay, 1838, 1, p. 57).

Middle- and upper middle-class travellers like the then healthy Macaulay in 1838 were thus both dependent on her for hotel recommendations and travel tips and enjoyed caricatures of the English farcically tied to her apron strings. She died in 1838, aged 76, at Milan while travelling home to England (DNB, 1909); Murray and Baedeker were at hand to pick up her legacy and direct continuing generations of cultural and valetudinarian travellers. Not least she had taught them her multiple explanation mark system (!) for grading artworks, which was taken up as the star system (*) (Bruce, 2012). She often travelled in 'manly garb' even as Jack Starke (Parsons, 2007). Maybe because she was a woman, perhaps more because she did not continue to appear in print posthumously as the 'brand' of a namesake publishing house, Mariana Starke has been overlooked compared to John Murray and Karl Baedeker.

John Murray

Murray's early Continental guides elaborate on the German spas, but he does describe Switzerland in 1838, mentioning if only incidentally its then one or two spa towns (like Baden, near Zurich 'resorted to by numerous visitors... chiefly natives of Switzerland' (Simmons, 1970, p. 17).

On the way to Germany, however, Murray's 1840 edition detailed an Excursion to Spa despite dismissing the resort 'as a watering-place [which] is much fallen off and its scenery very inferior to that of the Rhine' and 'in fact, out of fashion' (1840, p. 174). 'Since 1834 the English have deserted it for the Brunnen of Nassau, which far surpass Spa' (1840, p. 175), but he says in its time as 'the first watering-place in Europe monarchs were as plentiful as weavers from [the local town of] Verviers now are' (1840, p. 175).

On pages 217–20 of his work, Murray includes a general two and half thousand word essay on *GERMAN WATERING-PLACES*. 'For the Germans [of all classes] an excursion to a watering place in the summer is essential to existence.' Inter alia he describes how a [German] Government minister

> Repairs thither to refresh himself... but usually brings his portfolio... nor does he... altogether even here bid adieu to intrigue and politics. The invalid comes to recruit his strength – the debauchee to wash himself inside and out, and string his nerves for a fresh campaign of dissipation – the shopkeeper and the merchant come to spend their money and gaze on their betters, and even the sharper and black-leg, who swarm at all the baths, to enrich themselves at the gaming tables at the expense of their fellow guests.

Intrigue, dissipation and gambling feature much more strongly than cure among the attractions of places that are categorised more by their class of guest than cure – 'Carlsbad, Toeplitz, and Bruckenau are the resort of emperors and kings; Baden and Ems of grand dukes, princes and high nobility. Wiesbaden is a sort of Margate where the overflowing population of Frankfurt repairs on Sunday afternoon... ' One only, Schlangenbad is named and then rather disparagingly as 'frequented by those whose business it is to be cured, and who are strenuously endeavouring, by a few weeks of abstinence and exercise, to extricate themselves from doctors' bills and the sick list' (p. 218).

Murray himself quotes extensively from a contemporary pamphlet *Autumn near the Rhine* on gaiety as the essence of German watering-places:

> If ennui may be the motive of as many visits to Aix-la-Chapelle [Aachen], as to similar places in Great Britain, the remedy here appears more successful: for you can rarely read in a single countenance, as you may in the libraries of Brighton or Cheltenham, the inveterate disease of which persons come to be cured... The Grand Saloon...with the exception of much more gaiety, more avowed vice, and the absences of all pretence at rational resources, acts the part of the library at an English watering place. An anxious silence reigns, only interrupted by the rattling of the roulette, the jingling of Napoleons [gold coins] and francs....Pretty interesting women were...seeing them [their Napoleons] swept away, or drawing them in doubled, with a sang-froid which proved that they were not novices.
>
> (See p.15 for another compatible or consistent comment on Aachen.)

Dismal cure seeking was clearly not the reason to go further afield than Cheltenham to take the waters, but he goes on to say that visitors should go with friends as 'like an English watering place...there is little society found out of your own circle' (p. 220). As a distinct afterthought Murray adds that 'this work does not pretend to describe the medical properties and sanatory powers of the various mineral springs... [users] should consult their own physician before leaving home....[and it is] prudent and customary to ask...the physician resident at the baths as well' (p. 220).

Murray's guides were popular and were followed by, for example, the valetudinarian and culture seeker Ruskin, although he was later bitterly critical of them (Hanley and Walton, 2010), by Dickens travelling in 1844 (Flint, 1998), Macaulay in the 1850s (see below) and of course many others.

Karl Baedeker

Karl Baedeker, perceived as the giant of nineteenth-century guidebook writers (Bruce, 2010, p. 93) and publishers, even in English, was Murray's valued agent in Coblenz by 1836 although already publishing his own German and French guidebooks and a three-way language manual.

An 1878 edition of his *Traveller's Manual of Conversation*, which in its introduction aims to be 'Little and good', includes 85 words for maladies and infirmities, including the ague, plague, the Cholera, consumption, gout, the cancer, gangrene and leprosy (p. 46–9) before finally coming after 'epidemic – eine Herrschende Krankheit – une epidemie – una contagion, epidemia' to a further 'diet'; 'a remedy...a bath...bleeding...cupping glass...the leech...the cure...' (p. 50). Perhaps wisely, no attempt is made at a dialogue with a doctor although 'Engaging a servant' (p. 292) and 'To buy a travelling carriage' (p. 310) are usefully included.

By the 1900s, Baedeker's introduction to his handbook to Switzerland included seven pages on the 'Climate of Switzerland. Health Resorts', which categorised every altitude and resort in Switzerland by its suitability as a treatment centre (Baedeker, 1907, p. xix–xxvii). *Purity of the Atmosphere, Warmth, Decrease in Atmospheric Pressure* and *Moisture* are each found significant and closely related to *Height*. Although many conditions are reported as catered for, invalids with '*Pulmonary* and *Nervous Ailments* are by far the most numerous' (p. xxi). Among the pulmonary, *Phthisis* sufferers are especially recommended Davos in winter after the season's snow fall, 'free from both dust and wind' (p. xxii).

On the other hand, among *Nervous Patients*, it is recognised that not only climate should be considered. '*Neurasthenics* maybe driven frantic by brass bands, by the rattle of the nine-pin alley, or by other noisy amusements; and the effect of the grandest Alpine air may in this manner be frustrated.'

Baedeker ends by noting that many places in contrast 'have been developed in recent years as winter-resorts for sport' (p. xii) 'and at some of these consumptive patients are not received' (p. xxii). The table of resorts and resort hotels contains some seven or eight hundred places ranging in height in English feet (approx 0.3 metres) from 636 to 8495. Davos with its *Magic Mountain* (Lowe-Porter 1960) sanatoria is between 5000 and 6000.

Switzerland is only the most extreme example of Baedeker's commitment to valetudinarian as well as cultural tourism; *Northern Italy* (1899), for instance, has nearly a page (p. 400) on the Baths of (Bagni di) Lucca.

The Bagni di Lucca (the baths of Lucca)

Starke not only details the things invalid or valetudinarian travellers need to take with them – from 'a Man-servant' to 'pure opium', but also describes a number of spas in Italy in great detail. Among them, naturally as a favourite of the British is Bagni di Lucca; this collection of villages along a mountain river is some twenty kilometres out of Lucca itself in western Tuscany but then significantly in the independent Grand Duchy/Principality and former Republic of Lucca (Figure 2.1).

At the Congress of Vienna (1814) Bagni di Lucca, as part of the Duchy of Lucca, was assigned to the Bourbon Maria-Louisa, Sovereign of Parma (Starke, 1837, p. 48). It continued as a popular summer resort, particularly for the English, who had a Church there (still extant). In 1847 Lucca with

Figure 2.1 Bagni di Lucca, 1907, postcard

Bagni di Lucca was ceded to the Hapsburg Grand Duke of Tuscany, Leopold II of Lorraine (that is to say finally though briefly under the same domain as Florence (Baedeker, 1899, p. 395)). His rule started a period of decline and in 1853 the Casino seems to have been closed, though later reopened after Lucca was included in a unified Italy in 1861 (1899, p. 400).

Two successive ruling princesses of Elisa Bonaparte Baciocchi (sister of Napoleon) and then the Bourbon Maria Louisa patronised the Spa, built a summer Residence and formed the centre of a small court and society. Typically for a Spa there was as much gambling as curing, and to this day, it is claimed (with very little evidence and some clear contradictions) that roulette was invented there (Jepson et al., 2006 and on websites). That in 1840 there was one of the first legal Casinos in Europe is more likely as the ruler of Lucca was then amenable.

Lord Byron in 1822 took a house here as did Heinrich Heine in 1828; later came Elizabeth Barrett Browning, perhaps more earnestly in pursuit of relief from the summer heat of Florence, where she lived from 1846 until her premature death in 1861. She celebrated her eternal love for Robert Browning on the banks of the river – they visited regularly, latterly with their young son, between 1849 and 1857 (local plaques seen 2011) (Ward, 1967, p. 147).

Starke, Murray and Baedeker (and indeed the Rough Guide 2006) all describe and emphasise different aspects. The Casino is recently re-opened and the English Pharmacy is still there (Figure 2.2).

Figure 2.2 English Pharmacy, 2011, called 'Betti's' in Baedeker (1899, p. 400)

Nineteenth-century valetudinarian and invalid travellers

Using literary works, the *Illustrated London News* and other periodicals and newspapers of the time, as well as biographies of nineteenth-century figures, there is a wealth of documents to be researched on the progress of a European (mainly) bourgeoisie staving off deadly (and lesser) illnesses through travel to 'healthy' places in foreign parts. From Keats to Karl Marx, Tolstoy, Chekhov, Turgenev, Elisabeth Barrett and Robert Browning, the Lady Trevelyan, who patronised the Pre-Raphelites, Ruskin, Wilkie Collins, Chopin, Georges Sands, Charles Darwin: all travelled for their health and the list could well be extended.Pickering has dealt with some of them and some more from the questionable *Creative Malady* standpoint (Pickering, 1974). This chapter can only touch on a few.

Lady Trevelyan was taken abroad in extremis, preferring to 'die quickly doing something she enjoyed in the company of people she loved [her husband Walter and John Ruskin among a couple of others], rather than go on with the lingering pain in familiar and safe surroundings' (Batchelor, 2006, p. 229). Following a journey in a special carriage attached to a train to Neuchatel, she duly died in 1866.

As a traveller for his health, Thomas Mann, though straying more into the twentieth century, could feature his utterly pre-war fictional characters from *The Magic Mountain* (Lowe-Porter, 1960). Gustave Aschenbach in the original novella *Death in Venice* (1911) gave Thomas Mann great scope for describing 'Sickness, its stealthy onset and its muted threat [which] fascinated him' (Winston, 1982, p. 278). The Benjamin Britten version of the novella is seen as a classic of tourism-related opera (Botterill, 2012). The simultaneous illness of his wife gave him an insight into the life of Davos, the real Magic Mountain (Winston, 1982).

Ruskin's early family travels were modelled on a previous generation's Grand Tour, but travelling as a family unit, even if for an educational purpose (Hanley and Walton, 2010, p. 43), was in the spirit of his own times. He also went on travelling to protect what his mother saw as his delicate constitution.

Macaulay's recently published Journals – five volumes and about 1500 printed pages – give considerable detail of his holidays and travels – in the UK then including Ireland and on the Continent. Suffering from cardiac problems from, at latest, 1852, his health and medical treatment looms large, and spa resorts were frequently his destination. Doctors (and quacks avoided) and his activity at spas feature significantly. Macaulay denied ever being a 'valetudinarian', thinking perhaps disdainfully of Mr Woodhouse in *Emma* – Jane Austin was a favourite author of his.

He had visited Great Malvern, a notable watering-place in August 1851 and

> ... Went to Dr Wilson's ... [not as a patient]. The Dr lives in a huge boarding house, filled with patients whom he lodges, feeds, and provides with a handsome garden for exercise after duckings [sic]. He must make a good thing of it: for he is quite like Jephson of Malvern. Thank God, I want no medicine at all. In any case I should not go to a quack.
>
> (Macaulay, 1851, p. 147–8)

Henry Jephson (1798–1878) was 'the celebrated doctor of Leamington, who drew patrons from all over Great Britain and the Continent. His income is said to have been over £20,000 [say £1.6million!] a year for several years'. He was a doctor who had treated both Macaulay's parents (Thomas, 2008, 3, p. 148).

He did not always get on with spas. He had fled from London and Edinburgh politics to Clifton (Bristol), after his first severe heart incident in July 1852, and he wrote in August of 'vile drizzling weather – I shall never come to this place again. The scenery is lovely and the air pure; but the eternal rain is to me intolerable. I wish I were in Ventnor again or at Malvern' (Macaulay, 1852, p. 280), but he stayed another month, described many a pleasant walk, visits to places his mother's family had earlier inhabited[1] and did not complain again.

Late in 1859 his two distinguished London Doctors – 'Watson [later made a Baronet] and Martin, came to consult – They agreed in pronouncing my complaint a heart complaint only. If the heart acted with force all the plagues would vanish altogether, the phlegm, the coughing, etc etc. They are going to give me steel and quinine. But they seem to be strongly of opinion that I ought to try next winter some milder climate, possibly Madeira. They may be right' (Macaulay, 1859, p. 394). But they and he were too late. The steel powder – shades of Hooke's early (self-)experimentation with various substances including a head numbing 'wine mixed with steel filings' (Johnson, 2011, p. 102) two centuries earlier – was supposed to quicken circulation; the quinine was a tonic (Thomas, 2008, p. 395) – a couple of days later the patient made unknowingly his final journal entry: he records a terrific shock as a ceiling came down narrowly missing him on his water closet; he then typically went on to write at length about a literary controversy involving Dickens and Leigh Hunt. He died at home in his armchair within the week, aged 59.

Conclusions

The nineteenth century was, as we have seen, not a comfortable time to be sick. Spas and keeping away from doctors may have been the better of real evils so long as the individual had ample wealth. An industry of spa and health travel built up around that wealth, and Starke, Murray and Baedeker mediated the choices of travellers. If the bourgeois consumer was not too ill, spas afforded great pleasure and delights, preferably far from the prying eyes of family and social equals. Even nearer at hand they could provide pleasant surroundings and some relief of symptoms. As the American Grocer's Encyclopaedia put it: there were benefits to be had from 'drinking large quantities of innocuous liquid' (Ward, 1911, p. 387), and when the scientific approach was still pretty lethal, the largely placebo effect of 'taking the waters' was both safer, less painful and could even be fun.

Note

1. Selina Mills, an acolyte of Hannah More, married Zachary Macaulay, the Slave Trade Abolitionist in Bristol in 1799.

References

Ackroyd, P. (2012) *Wilkie Collins: A Brief Life* (London: Chatto and Windus).

Baedeker, K. (1878) *Traveller's Manual of Conversation in Four Languages*, 23rd edn (Leipsic: Karl Baedeker).

Baedeker, K. (1899) *Northern Italy*, 11th remodelled edn (Leipsic: Karl Baedeker).

Baedeker, K. (1907) *Switzerland*, 22nd edn (Leipsic: Karl Baedeker).

Batchelor, J. (2006) *Lady Trevelyan and the Pre-Raphaelite Brotherhood* (London: Chatto and Windus).

Botterill, D. (2012) 'Representations of tourism in 20th century opera' in J. Tivers and T. Rakic (eds) *Narratives of Travel and Tourism* (Farnham, Surrey: Ashgate).

Bowring, E. A. (1860) *Return of the Number and Nature of the Accidents and the Injuries to Life and Limb Which Have Occurred on All the Railways Open for Traffic in England and Wales, Scotland, and Ireland respectively,* from the 1st January to the 30th June 1860 (Command Paper HMSO).

Bruce, D. M. (2010) 'Baedeker: The perceived inventor of the formal guidebook – a bible for travellers in the 19th Century' in R. Butler and R. Russell (eds) *Giants of Tourism* (Wallingford: CABI).

Bruce, D. M. (2012) 'The nineteenth century "golden age" of cultural tourism' in G. Richards and M. Smith (eds) *Handbook of Cultural Tourism* (London: Taylor and Francis).

Butler, R. and Russell, R. (eds) (2010) *Giants of Tourism* (Wallingford: CABI).

Buzard, J. (1993) *The Beaten Track: European Tourism, Literature, and the Ways to Culture, 1800–1918* (Oxford: Clarendon Press).DNB (1909) 'Starke, Mariana (1762–1838) CFS (Miss C. Fell Smith)' in *Dictionary of National Biography* Vol. XVIII, 994 (Oxford: OUP).

Dunning, E. (1996) 'Problems of the emotions in sport and leisure: critical and counter critical comments on the conventional and figurational sociologies of sport and leisure' *Leisure Studies*, 15, 185–207.

Flint, K. (1998) 'Introduction' in K. Flint (ed.) *Pictures from Italy* by Charles Dickens (1846) (London: Penguin).

Fuessel, S. (ed.) (2008) *Georg Braun and Franz Hogenberg – Cities of the World 1572–1618* (Koeln, Germany: Taschen).

Gittings, R. (1968) *John Keats* (London: Penguin).

Hanley, K. and Walton, J. (2010) *Constructing Cultural Tourism: John Ruskin and the Tourist Gaze* (Bristol: Channel View).

Illustrated London News for 29 May (1847) (Chester) [for one of many examples].

Jepson, T., Buckley, J., and Ellingham, M. (2006) *The Rough Guide to Tuscany and Umbria* 6th edn (London: Rough Guides).

Johnson, B. (2011) *Johnson's Life of London* (London: Harper)

Koshar, R. (2002) 'Introduction' in R. Koshar (ed.) *Histories of Leisure* (Oxford: Berg)

Levi, P. (1971) *Pausanias Guide to Greece 2nd Century AD*, translated with an Introduction 2 Volumes (London: Penguin)

Macaulay, T. B. (1838–9; 1848–1859) 'The Journals of Thomas Babington Macaulay' in W. Thomas (ed.) (2008) (London: Pickering and Chatto).

Magner, L. N. (2007) *A History of Medicine* (New York: Informa Healthcare Inc).

Mann, T. (1960) *Der Zauberberg* (The Magic Mountain), translated by H. T. Lowe-Porter, (London: Penguin).

Mullen, R. and Monson, J. (2009) *The Smell of the Continent: the British Discover Europe* (London: Macmillan).

Murray, J. III (1840) *Handbook for Travellers on the Continent: Being a Guide through Holland, Belgium, Prussia and Northern Germany etc.* (London: John Murray and Son).

Oxford (1973) *The Shorter English Dictionary on Historical Principles* (Oxford: Oxford University Press).

Parr, S. (2011) *The Story of Swimming* (Stockport: Dewi Lewis).

Parsons, N. T. (2007) *Worth the Detour: A History of the Guidebook* (Stroud: Sutton Publishing).

Perrottet, A. (2012) 'Essay: Guidebooks to Babylon' *New York Times Review* 20 January 2012.

Pickering, G. (1974) *Creative Malady: Illness in the Lives and Minds of Charles Darwin, Florence Nightingale, Mary Baker Eddy, Sigmund Freud, Marcel Proust, Elizabeth Barrett Browning* (London: Allen and Unwin).

Simmons, J. (ed.) (1970) *Murray's Handbook for Travellers in Switzerland 1838, Facsimile Edition and Introduction* (Leicester: Leicester University Press).
Starke, M. (1800) *Letters from Italy between the Years 1792 and 1798 etc in two Volumes* (London: Phillips).
Starke, M. (1837) *Travels in Europe for the Use of Travellers on the Continent Including the Island of Sicily* 9th edn (London: John Murray).
Stendahl (1839/2006) *La Chartreuse de Parme* (The Charterhouse of Parma) translated by J. Sturrock (London: Penguin).
Sterne, L. (1764/1911) *A Sentimental Journey through France and Italy* (Toronto: William Briggs).
Thomas, W. (ed.) (2008) *The Journals of Thomas Babington Macaulay Five Volumes* edited with introduction and notes (London: Pickering and Chatto).
Towner, J. (1995) 'Tourism's Histories' *Tourism Management* 16, 5, 339–43.
Towner, J. (2010) 'The Master of Ceremonies: Beau Nash and the Rise of Bath' in R. Butler and R. Russell (eds) *Giants of Tourism* (Wallingford: CABI).
Trease, G. (1967) *The Grand Tour* (London: Heinemann).
Unsigned (1911) 'Mineral Waters' *Encyclopaedia Britannica* 18, 517–22.
Urry, J. and Larsen, J. (2011) *The Tourist Gaze 3.0* (London: Sage).
Walton, J. (2010) 'Thomas Cook: Image and Reality' in R. Butler and R. Russell (eds) *Giants of Tourism* (Wallingford: CABI).
Ward, A. (compiler) (1911) 'Mineral Waters: 387–391' *Encyclopedia of Foods and Beverages* (New York: Grocer's Encyclopedia).
Ward, M. (1967) *Robert Browning and His World*, 1, (London: Cassell).
Wikipedia (2012) 'Aachen: The Middle Ages', en.wikipedia.org/wiki/Aachen, date accessed 19 March 2012.
Winston, R. (1982) *Thomas Mann: The Making of the Artist, 1875–1911; From His Childhood to the Writing of Death in Venice* (London: Constable).

3

A Way Through the Maze: Exploring Differences and Overlaps Between Wellness and Medical Tourism Providers

Cornelia Voigt and Jennifer H. Laing

The aims of this chapter are as follows:

- to develop a typology of medical and wellness tourism providers and to explore differences and overlaps between the two forms of tourism;
- to examine five provider models where the lines between medical and wellness tourism have been blurred;
- to compare the concept of wellness with the biomedical conception of health;
- to provide background information on the broader environment of health care and Western consumer culture, both of which have influenced or played a role in the development of these provider models;
- to discuss areas for future research.

Introduction

The literature about the concepts of medical and wellness tourism lacks consistency. While some suggest that medical tourism subsumes the concept of wellness tourism (for example, TRAM, 2006), or vice versa (for example, Sheldon and Bushell, 2009), others refer to the theoretical difference between the 'illness' concept and the 'wellness' concept, in order to meaningfully differentiate between contrasting health tourism products and services. The first aim of this chapter is to provide a typology of medical and wellness tourism as two distinct segments of health tourism.

However, this chapter acknowledges that there are overlaps between medical and wellness tourism offerings. In recent years, several business models have emerged, based on diverse and unique synergies between biomedical health care, and the wellness and hospitality industries. While we argue that medical and wellness tourism are fundamentally distinct, there are enough similarities in the actual provider business models to warrant the inclusion of wellness tourism in a book about medical tourism. To date, these overlaps have been largely ignored in the tourism and health sciences literature. Using short vignettes, the second aim of this chapter is to outline five provider models where the difference between 'illness' and 'wellness' has become increasingly blurred: (1) health clusters, (2) 'wellspitals', (3) therapeutic lifestyle retreats, (4) 'medhotels' and (5) medical spas. The emergence of these models needs to be understood in the context of broader changes in health care and Western consumer culture. Different cultural developments in the global health tourism supply structure also need to be taken into account. We therefore briefly examine philosophically divergent views of health, the rise of complementary and alternative medicine (CAM) and integrative medicine (IM), the increased privatisation of global health care and the recent links between health, consumerism and beauty. This informs our analysis of the similarities and differences between 'pure' medical and wellness providers, as well as the five 'in-between' types of providers.

Wellness as an alternative view to biomedical conceptions of health

'Wellness' is a complex construct with widely diverging definitions and different cultural and linguistic meanings attached to it (Smith and Puczkó, 2008). It has been defined as a lifestyle (for example, Hattie et al., 2004), a process (for example, Travis and Ryan, 1988) and a positive psychological state (for example, Ryan and Deci, 2001). It has also been described as an alternative to the biomedical conception of health.

The biomedical health paradigm equates health with the absence of disease and is associated with the Cartesian separation of mind and body. A prominent metaphor of the biomedical health model compares the body to a machine, in which malfunctioning body parts are seen as independently curable or replaceable (O'Connor, 2000). Conventional medicine is reductionist because it conceptualises illness exclusively as a biological process, where specific pathogens damage the 'body machine' either acutely, chronically or fatally. Broken machines do not repair themselves from the inside; they need an expert to diagnose the problem externally. Accordingly, biomedical health professionals attempt to find deviations in a patient's body structure, often with increasingly sophisticated high-tech diagnostic tools (Davis-Floyd, 2001). On the basis of an often technologically intermediated diagnosis, the patient is treated, usually involving pharmaceutical drugs and/or surgery. During this process, practitioners tend to objectify patients and interact with them in an emotionally detached manner. The doctor–patient relationship can be described as paternalistic, where the doctor has the authority and control of information, which frees the patient of responsibility (Davis-Floyd, 2001).

In the holistic wellness paradigm, health is viewed as a positive state, namely the presence of well-being, rather than the absence of illness. Instead of focusing on pathogens, the wellness paradigm takes into account the 'whole' person by acknowledging the interdependence of body, mind and spirit. Dunn (1959, p. 447) emphasises that there is no optimum level of wellness. Instead, each individual strives towards an 'ever-higher potential of functioning' of what they are capable of (a view that supports the process idea of wellness). Consequently, wellness is a subjective concept. Individuals, who might be considered ill by biomedical standards, can still perceive themselves as well. Accordingly, even individuals who are chronically or terminally ill are able to strive towards higher levels of wellness, thus demonstrating health improvement in a broad sense. Conversely, people without symptoms of physical or psychological illness might still be 'unwell'.

In the wellness paradigm, the focus is not so much on the treatment of symptoms, but on self-healing, health promotion and illness prevention. This view of health implies a proactive, personal commitment to strive for well-being. In the majority of definitions of wellness, these elements of *individual action*, *self-care* and *self-responsibility* are evident. Wellness practitioners are aware that their role is to support people to evoke their self-healing capacities. They treat from the inside out by trusting the inner recognition of their clients and individual self-healing capacities, as well as their own intuition (Davis-Floyd, 2001). In the wellness paradigm, therapeutic encounters are characterised as subjective and are based on welcoming individual experiences and emotions, as well as being collaborative, participatory and 'low-tech/high touch'. Rather than being compliant and deferential, patients become active, empowered and informed agents in relation to their own health care.

CAM and IM

Since the 1960s, the holistic health movement has taken up the cause of wellness principles, which has influenced the growth of complementary and alternative medicine (CAM). This encompasses a variety of modalities (for example, massage, meditation) as well as entire alternative medical systems from Western (for example, homeopathy, naturopathy) and non-Western cultures (for example, Traditional Chinese Medicine, Ayurveda) that fall outside the boundaries of orthodox biomedicine. Despite the diverse cultures and beliefs on which these modalities and systems are based, they share a conceptual core based on the wellness paradigm. Shared principles include the following: (1) healing the whole person rather than curing specific physical symptoms; (2) an integrated view of body/mind/spirit; (3) a focus on natural, non-invasive interventions and botanical medications; (4) the belief in the innate ability of the human organism to heal; (5) the importance of self-care; and (6) patient empowerment (Goldstein, 2000; O'Connor, 2000; Kelner and Wellman, 2003).

CAM is associated with or even defined as a counter-cultural movement that has challenged the dominance of biomedicine (Kelner and Wellman, 2003; Saks, 2003). Some argue that consumers have been largely responsible for the growing acceptance of CAM (Kelner and Wellman, 1997; Eisenberg et al., 1998). According to this discourse, 'consumers' can be understood as political agents who want to challenge the status quo through the exercise of their consumer power, and 'consumerism' is seen as a social movement of protest and resistance (Williamson, 1999).

Neither orthodox health professionals nor health educators and governments have been able to ignore the growing uptake of CAM and consumer dissatisfaction with conventional medicine. This has led to the inclusion of CAM and holistic health subjects into the curricula of a number of biomedical faculties of leading universities (Baer and Coulter, 2009). More governments are prepared to fund CAM research, and there is wider knowledge dissemination through a growing number of academic journals and monographs dealing specifically with CAM. The focus is moving from CAM to *integrative medicine* (IM), which aspires to blend conventional biomedicine and CAM to the greatest benefit of individual patients.

As with many new developments in health care, the development of IM has caused heated debates and become a highly charged political issue. There are proponents, often mainstream medical professionals, who have enthusiastically embraced IM and view it as the start of a genuine paradigm shift in medicine (Gaudet and Faass, 2001; Dobos, 2009) or at least a vision of what future health care could become (Bodeker and Kronenberg, 2002; Snyderman and Weil, 2002). Many CAM practitioners have developed professionalisation strategies to increase legitimacy and acceptance of CAM by mainstream medicine, including the formation of professional associations and development of codes of conducts and standardised educational programmes, as well as the search for recognition from an accrediting or certifying agency and state regulation (Saks, 2000, 2003). Some orthodox health professionals however still reject CAM outright, let alone the integration of it with biomedicine. Dr Arnold Relman, Professor Emeritus of Medicine and Social Medicine at Harvard, posits that integrating CAM with biomedicine 'would not be an advance but a return to the past, an interruption of the remarkable progress achieved by science-based medicine' (Dalen, 1999, p. 2122). Others fear that IM is 'a smokescreen for smuggling unproven treatments into the realm of rational medicine' (Ernst, 2011, p. 1).

Some have doubts about IM because they fear that there has been a takeover of CAM by the orthodox medical community to pursue their goals of controlling and avoiding competition, as well as utilising CAM for profit-making ventures (Baer, 2002, 2008; Hollenberg, 2007; Baer and Coulter,

2009). The possibility of making profit out of CAM integration must be viewed in light of the increasing global trend of privatisation and corporatisation of health care. For private, for-profit health-care providers, rising customer demand for CAM, coupled with the fact that the majority of people pay CAM treatments out of pocket or by private health insurance, may present a tempting business opportunity (Baer, 2008). The previously discussed collective demands for political rights and freedom of choice become re-interpreted as individual consumer desires for value for money and customer satisfaction from an economic perspective. In this sense, CAM has become a commodity from which to generate profits (Hollenberg, 2007), while IM offers an opportunity to expand market share in a competitive environment (Kaptchuk and Miller, 2005). However, as CAM therapies are usually not covered by public health care, consumer choice is available only to those who can afford it. Consequently, 'consumerism' takes on an ideological meaning, with the purpose of legitimising capitalism (Shaw and Aldridge, 2003).

Despite these very different views, medical pluralism is a reality. A great majority of CAM users do not perceive CAM as a substitute for biomedicine but employ a pragmatic mix and match of biomedical and CAM treatments according to their needs (Kelner and Wellman, 1997; Eisenberg et al., 1998), but without necessarily taking into account the comparative maturity of evaluation of the proven effectiveness of both approaches.

Biomedicine, CAM, IM, privatisation of health care and lifestyle are all important elements underpinning the concepts of medical and wellness tourism. The final element that needs to be discussed is beauty.

Beauty

While the notion of 'beauty' does not appear to a large extent in conceptualisations about health or CAM, it has been included in *tourism* models of wellness as a defining component of physical wellness (Müller and Lanz Kaufmann, 2001). It is also argued that health promotion messages about healthy eating and exercising have been overlaid by commercial interests, which overemphasise the aesthetic 'by-product rewards' resulting from such a lifestyle (Featherstone, 1991). Mass media and the fashion, cosmetic and fitness industries are reinforcing and capitalising on this beauty ideal. The body 'becomes a site of different choices and different possibilities' and people turn into 'designers of their own bodies' (Askegaard et al., 2002, p. 800) – ranging from applying an anti-wrinkle cream to teeth whitening and even more invasive practices such as breast augmentations.

In the areas of cosmetic surgery and aesthetic medicine (that is, minimally invasive cosmetic procedures such as botox injections), the commercialisation of beauty is linked to biomedicine. Like CAM, cosmetic surgery has grown rapidly in the past 30 years and has become a more accepted practice of self-maintenance (Gimlin, 2000). Sophisticated medical techniques have allowed unprecedented body modification for non-medical, aesthetic reasons. Women, and increasingly men, use cosmetic surgery as a 'lifestyle choice', acquiring a 'remedy' for surplus weight, stereotypical markers of ethnicity, age and other perceived flaws (Gimlin, 2000; Askegaard et al., 2002; Ackerman, 2010). However, for some, there still seems to be qualitative differences between applying an anti-wrinkle cream and making permanent changes to the body.

Achieving wellness through living a healthy lifestyle requires continuous effort and discipline and involves 'rejecting things that the dominant consumer culture presents as good: an excess of materialism, hedonism, and impulsiveness' (Goldstein, 2000, p. 31). The consumer culture's rhetoric of taking responsibility for one's life however excludes restraint and physical exertion.

These broader developments and tensions in health-care delivery and consumer choices help to explain the differences and overlaps between different types of medical and wellness tourism providers. We introduce a typology of medical and wellness tourism providers as a starting point for analysis.

Exploring differences and overlaps between medical and wellness tourism

A typology of health tourism

As mentioned earlier, there are no consistent and widely accepted definitions of terms such as 'health tourism', 'medical tourism' and 'wellness tourism', which are often used interchangeably. We suggest that wellness and medical tourism be considered as two distinct segments of health tourism, based on the theoretical difference between 'illness' and 'wellness' concepts, and the two different underlying health paradigms (Figure 3.1; Voigt et al., 2010). This theoretical distinction can be used to meaningfully differentiate the vast diversity in supply (that is, facilities, products, services) or demand (that is, different motives, expectations, experiences) (Müller and Lanz Kaufmann, 2001; Puczkó and Bacharov, 2006). Health tourism is used as a comprehensive umbrella term that subsumes both medical and wellness tourism.

Medical tourism providers are 'illness-oriented'. They cater to tourists who primarily travel in order to cure or treat a certain disease or medical condition, which also includes the desire for a permanent change in physical appearance. The majority of services offered by medical tourism providers are general biomedical procedures, including invasive and high-tech diagnostic services. Employees treating medical tourists are typically orthodox health professionals, such as doctors and nurses.

In contrast, wellness tourism providers are 'wellness-oriented'. They cater to tourists who primarily travel to maintain or improve their health and

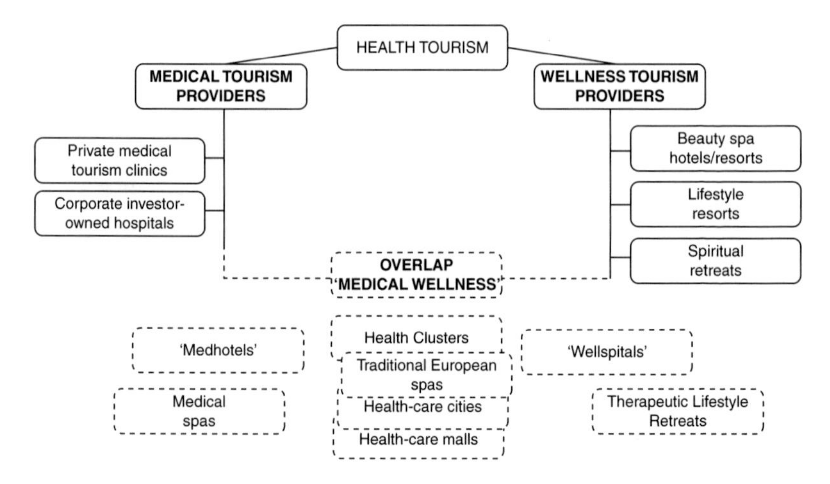

Figure 3.1 Typology of medical and wellness tourism providers

well-being. Services offered by wellness tourism providers generally fall outside the realm of biomedicine. They encompass a range of services, including CAM therapies, non-medical beauty treatments, and spiritual and lifestyle-based interventions. Employees treating wellness tourists are normally not mainstream health professionals, such as CAM practitioners, beauticians, nutritionists and lifestyle coaches. It has been suggested that while medical tourists want to become healthier, wellness tourists already perceive themselves to be healthy (Müller and Lanz Kaufmann, 2001; Verschuren, 2004). They might have minor health problems such as insomnia or back pain. However, rather than turning to biomedical solutions to deal with these issues, wellness tourists search out holistic methods to achieve higher levels of wellness.

Medical tourism providers

'Pure' medical tourism providers include private medical tourism clinics and corporate investor-owned hospitals or hospital chains. An example of a chain is the Indian-based *Apollo Hospitals*, which provide a vast range of invasive and diagnostic services. While Bookman and Bookman (2007) argue that diagnostic tests are increasingly used for health prevention, they are geared towards illness detection and physical abnormalities. The difference between conventional private hospitals and private hospitals targeting medical tourists is that the latter have successfully turned themselves into 'hotelspitals', specifically designed to make foreigners more comfortable. This is achieved through luxurious décor and facilities, and high-attentive service personnel, similar to high-class hotels (Cohen, 2008a). Smaller, private clinics tend to focus on a specific area of biomedical procedure, predominantly offering elective treatments which are undertaken for non-medical reasons (for example, cosmetic surgery, dentistry or in vitro fertilisation). Examples include the *Dentourist Clinic* in Hungary and the *Barbados Fertility Centre*.

Wellness tourism providers

Wellness providers do not represent a homogenous group and there are considerable differences in the facilities and services offered. While spas form the largest segment, there are two other types: lifestyle resorts and spiritual retreats (Voigt et al., 2010, 2011). Research has shown that there are statistically significant differences between the socio-demographic profile, travel behaviour and travel motivations of visitors to these different types of providers (Voigt et al., 2011). What these providers have in common is their holistic orientation to health, a focus on health promotion and the provision of CAM procedures.

The main focus of *beauty spa hotels/resorts* is on body and non-medical beauty treatments such as massages, body wraps and facials. Some beauty spa providers have an intrinsic geographic advantage as they are based on or around mineral or thermal pools. Spas that do not have this advantage can offer water-based treatments, in the form of steam rooms, saunas, whirlpools, hot tubs and wet treatment rooms. Massage and hydrotherapy are considered to be CAM treatments. Day spas are similar to beauty spa hotels/resorts, but do not offer accommodation. Examples are the *Banyan Tree Club and Spa* in Korea and the *Peppers Mineral Springs Retreat* in Australia.

Beauty spa visitors are passive recipients of treatments from a therapist. In contrast, *lifestyle resort* visitors actively participate in a comprehensive programme, which focuses on health-promoting behaviour changes in areas such as nutrition, exercise and fitness, personal goals, and stress management. Most lifestyle resorts offer a vast range of CAM treatments and some programmes are entirely based on alternative health philosophies such as the *Siddhalepa Ayurveda Health Resort* in Sri Lanka. The cuisine at lifestyle resorts is very healthy and is often based on organic produce. Amenities that are normally part of every hotel or resort room, such as televisions and telephones, are generally not provided, and the consumption of alcohol, coffee and cigarettes is usually prohibited. Most lifestyle resorts have a beauty spa where treatments can be booked at an additional cost, but the focus on beauty is not as pronounced as in beauty spa hotels/resorts. Examples are the *Gwinganna Lifestyle Retreat* in Australia and the *Rancho la Puerta* in Mexico.

The emphasis of *spiritual retreats* is on spiritual development. They can be religious or non-religious but always include meditation in various forms. Many are based on some specific teachings or philosophy and/or focus on the study of a specific mind/body technique, such as yoga or T'ai Chi, or particular forms of meditation (for example, Transcendental Meditation, Vipassana). Mind/body techniques are generally considered to be part of CAM. Some of these retreats are 'silent', so participants can focus on their private spiritual journey. Facilities tend to be basic and visitors may have to share austere rooms and/or bathroom facilities. The cuisine is often vegetarian or vegan and alcohol is generally not offered. Examples include the *Dechen Chöling Shambhala Buddhist Meditation Center* in France and the *Yoga Haus Samvit* in Germany.

In addition to 'pure' medical and wellness tourism providers, there are other providers (*medical wellness* providers) that offer a greater fusion of medical and wellness services. Figure 3.1 depicts the five types of medical wellness tourism providers, outlined below.

Medical wellness providers

It is not clear when and where the term 'medical wellness' originated. Based on the preceding discussion of the contrast between the biomedical and wellness health paradigms, the term appears to make little conceptual sense. However, it symbolises a new coalescence between biomedical and CAM health professionals, and a stronger focus on wellness principles. In the US, 'medical wellness' is clearly linked to an IM approach to health care.

In Europe, the term 'medical wellness' is more directly linked to the wellness tourism industry. The different cultural meanings of the terms 'spa' and 'wellness' come into play. The term 'spa' in Europe traditionally implies medicine and cures (Puczkó and Bacharov, 2006). The more leisure-oriented concept has also arrived in continental Europe. What we have described here as 'beauty spa hotels/resorts' are usually referred to as 'wellness hotels' in continental Europe. There, the term 'wellness' is often used in marketing to invoke an image of passive pampering. The addition of 'medical' to 'medical wellness' is thus an attempt to re-emphasise active health promotion and perhaps sounds catchier than 'integrative medicine'.

Health clusters

The first example of a health cluster is the *traditional European spa*, which does not maintain a rigid separation between biomedicine and CAM. In the eighteenth century, they were seen as healing centres catering to people with health problems (Lempa, 2008), although also frequented by people seeking recreation instead of recuperation (Lempa, 2008). From the late nineteenth century, the healing aspects began to outweigh the emphasis on amusement and entertainment. There was a growing scientific analysis of the curative effects of water, as well as research on mud and seaweed treatments (Weisz, 2001). Entire townships, clinics and hospitality industries sprang up around thermal springs and sea coastlines, while spa visitation became part of mainstream medicine and was covered under public health insurance schemes. By the late 1990s, governments reduced their support for spa visitation, forcing spas to develop offers for self-paying guests and focus on illness prevention (Nahrstedt, 2004). A re-positioning and product diversification in traditional spa towns has taken place, in order to attract both medical and wellness tourists.

Bad Pyrmont in Germany is a traditional European spa town with a 400-year-old history of 'taking the waters'. Here, one single operator controls different medical and wellness facilities, such as a therapeutic health centre, a clinic which specialises in rehabilitation and pain relief, and a huge thermal bathing landscape, including pools, steam baths and saunas. Most health clusters in spa towns however involve a network of different organisations (for example, clinics, hotels, spa providers, tourism associations), who essentially collaborate in offering services to guests.

The two other types of 'health clusters' are *health-care cities* and *health-care malls*. The *Dubai Healthcare City*, launched in 2002, had the vision to become a regional hub, providing first-class medical *and* wellness services for treatment and prevention. Although it incorporates two hospitals and over 90 outpatient clinics and diagnostic laboratories, its health-care provider list reveals very few CAM or other wellness service providers. Thus far, it has failed to capture the medical tourism market (McGinley, 2011). Another health cluster where the wellness aspect is more prominent is the *Bangkok Mediplex*, a four-storey 'medilifestyle mall'. Biomedical and CAM treatment service providers represent about 70 per cent of the occupants. Whereas in some spa towns, biomedicine and CAM are truly integrated, services in the health-care city and medical mall models seem to just co-exist. It is not clear whether biomedical staff cooperate with CAM practitioners.

'Wellspitals'

On a smaller scale than 'health clusters', hospitals have started to specifically attract healthy people. An example is the *Klinik am Haussee* in Feldberg, Germany. The clinic mainly accommodates chronically ill people, but they have also established 'medical wellness' programmes that are responsible for 10 per cent of the company's revenue (Menzel, 2011). These target people with health problems (for example, those who are overweight or soldiers from crisis areas who are in need of recuperation), the 'worried well' (for example, those who want a comprehensive diagnostic health check-up to detect and prevent health problems) and guests who just want to relax through massages and spa treatments. Many of the latter are relatives who accompany a chronically ill patient.

Another example of 'wellspitals' are providers who specialise in the rehabilitation of people with psychiatric conditions (for example, anorexia, depression) and addiction. Again, integration of biomedicine, CAM therapies and health education helps people to detox, bolsters their self-esteem and self-care abilities and gives them a more positive outlook on life. 'Wellspitals' focusing on addiction might be well positioned to attract foreign patients, as travelling to another country facilitates anonymity and confidentiality (Connell, 2006). An example is the *Bhavana Phuket Residential Addiction Treatment Center* in Thailand, which employs Eastern- and Western-integrated methods, such as mindfulness meditation, acupuncture and journaling, to help people with addiction issues. Like medical tourism hospitals, 'wellspitals' emphasise their hotel-like premises and amenities. They frequently promote the beauty of the surrounding environment and both the *Klinik am Haussee* and the *Bhavana Phuket* highlight a range of natural and cultural tourist attractions in the vicinity.

Therapeutic lifestyle retreats

Therapeutic lifestyle retreats are similar to 'wellspitals', in that the lifestyle aspect is more pronounced. Unlike lifestyle resorts, they have been mainly developed for people affected by serious illnesses. An example is *The Gawler Foundation* in Australia which offers a broad range of one-day, weekend and residential programmes primarily for people with cancer. It is a non-profit organisation and less costly than the usual upscale lifestyle resorts. The Foundation advocates a self-help programme, encouraging and empowering individuals to actively work towards their healing and establish a healthy lifestyle. They see their programmes as a valuable augmentation rather than replacement of biomedicine.

'Medhotels'

'Medhotels' represent a collaboration between a private, biomedical clinic and a hotel to attract both wellness and medical tourists. They focus on medical tourists with non-permanent health problems (for example, burnout or back pain). Unlike lifestyle resorts, guests are not enrolled in a comprehensive programme covering a range of health-promoting areas. The emphasis tends to be on biomedical and high-tech diagnostic modalities. The partner hotels are often upscale with modern, luxurious facilities as well as beauty, fitness and spa offers. Examples include the *Alpenmedhotel Lamm* in Austria, which mainly concentrates on physiotherapy and mobilisation, and the *Klinik und Hotel St. Wolfgang* in Germany, which provides cardiology, urology and sports medicine services. Both providers also specialise in treating golf players, dealing with sport injuries and offering specific physiotherapeutic exercises to improve a patient's game.

Medical spas or 'Medispas'

Medical spas are another hybrid between biomedicine and the spa world. They can be day spas or built like a resort, including accommodation. Their main focus is on beauty and anti-ageing. To ensure quality and safety for the patients, they are supposed to be supervised by a physician. The aesthetic services offered in medical spas are not as invasive as cosmetic surgery but still pose greater health risks than conventional facials or body treatments. Examples of typical treatments are botox filler injections and aggressive chemical peels. Medical spa treatments tend to be expensive because they are dependent on costly technology. Some medical spas like the *Blue Water Spa* in Northern California, USA, are attached to the practice of an individual plastic surgeon. There is a continuum, with providers who solely offer aesthetic and cosmetic treatments at one end, those who provide a limited range of other spa treatments such as massages and larger providers in the middle and those who combine a focus on beauty and anti-ageing with additional biomedical services such as plastic surgery and diagnostics/services that can also be found in lifestyle resorts, such as nutrition counselling and exercise training at the other end. An example of a more comprehensive medical spa is the *Clinique La Prairie* in Switzerland, whose primary goal is 'to fight the effects of aging, while improving your quality of life' (Clinique La Prairie, 2011).

Coalescence between wellness tourism providers and mainstream health professionals

Some types of wellness tourism providers, specifically the spa industry, are in the early stages of professionalisation in order to become more accepted by mainstream health professionals and can be seen as an important element in future global health care. Since 2000, there has been a large increase in the establishment of international, national and regional spa and wellness tourism associations (Voigt et al., 2010). The rapid growth of the wellness industry has led to the need for comprehensive accreditation standards. Some associations have started to establish codes of ethics and robust accreditation schemes to assess the quality of wellness providers. Further evidence of professionalisation is the rising number of specialised diplomas, bachelor and master degrees such as the BA and MA of Health Management in Tourism at the *Joanneum University of Applied Sciences* in Austria and the MA of Wellness at *RMIT University* in Australia.

Some proponents of IM enthusiastically embrace spas as a potential tool for revolutionising global health care through an emphasis on wellness principles and illness prevention (Carmona, 2008; Cohen, 2008b). However, while many wellness tourism providers believe that they play a crucial role in health promotion and illness prevention, they feel that they are not taken seriously by mainstream medicine and health policy makers (Voigt et al., 2010, p. 91).

There also some cautious voices when it comes to synergies between wellness tourism providers and biomedicine. As Ellis notes: 'Spas might lose the caring touch that they are known for as they are influenced by the medical arena in which high tech trumps high touch resulting in significant depersonalization' (2008, p. 67). Furthermore, spiritual retreats and more alternative lifestyle resorts are rarely involved in professionalisation strategies and have no formal body to represent them. They may not even perceive themselves to be part of the tourism industry (Kelly, 2010).

Discussion and future research

The purpose of typologies is to provide a classification to deal with a chaotic situation such as the supply structure of health tourism providers. The preceding sections have examined the most 'typical' categories of medical and wellness tourism providers, classifying them based on the theoretical difference between 'illness' and 'wellness' and the two underlying health paradigms associated with each. Whether or not the providers consciously perceive this, 'pure' forms of medical and wellness providers clearly represent either the biomedical or the wellness paradigm of health.

However, developments in the health-care sector have increasingly led to an erosion of the boundaries between 'pure' forms of medical and wellness tourism. The rise of CAM, increasing dissatisfaction of consumers with the traditional paternalistic and technocratic doctor–patient relationship and a growing interdisciplinary research base showing that many non-biological processes influence people's objective and perceived health status have led to greater efforts to provide patient-centred optimal healing environments. These changes have created a conducive environment for introducing new business models in the health tourism arena and different types of 'medical tourism' providers have emerged, most only very recently.

Only traditional European spa clusters and 'wellspitals' appear to truly adopt an integrated approach that combines biomedicine with CAM and that has a strong focus on wellness principles. Rehabilitation of chronically ill people concentrates on (re-)learning self-care, while rehabilitation of people with psychiatric disorders and addictions focuses on a change of lifestyle habits and strengthening personal resources. This meshes well with wellness principles of self-responsibility, health promotion, patient empowerment and body/mind/spirit integration. Therapeutic lifestyle retreats also share these principles but do not combine their services with biomedical treatments and focus exclusively on CAM and lifestyle-based interventions. All these medical wellness providers cater to individuals considered to be ill by biomedical standards but who have the dual goals of illness treatment and health improvement. At the same time, many of these providers have diversified their products to cater to wellness tourists in the classic sense. More research is needed as to why some guests prefer medical wellness providers over 'pure' wellness tourism providers to meet their needs and to explore whether there might be problems inherent in mixing 'sick' and healthy tourist markets.

The spa sector in particular seeks to collaborate with mainstream health professionals and is more active in attempting to gain legitimacy and acceptance by biomedical health professionals through a number of professionalisation strategies. Spiritual retreats and smaller-scale lifestyle resorts on the other hand are not involved in either collaboration or professionalisation attempts. Future research needs to determine whether there is little shared interest in collaboration or whether they perceive themselves as 'counter-culturalists' who want to change the system rather than being co-opted by it.

'Medhotels', and to some extent traditional European spa and health-care city clusters, represent interesting scenarios where the biomedical community is collaborating with the hospitality industry (Ackerman, 2010). These constellations build fruitful ground for case study research about the benefits, challenges and costs of strategic alliances between organisations from two formerly separate industry sectors.

Most health tourism is almost exclusively driven by the private sector. As health is considered a public good in most countries of the world, the privatisation of health care is a hotly debated and politically charged topic. More research across all health tourism providers is needed to analyse the extent to which the idea of wellness is solely utilised as marketing rhetoric and to examine whether managers and staff actually know and embrace the wellness paradigm and its underlying principles.

Perhaps the largest area for future research relates to the assessment of the health effectiveness of services delivered by all health tourism providers. In the past two decades there has been a phenomenal growth in the development of rigorous subjective health measures that assess previously elusive, positive concepts such as psychological well-being, quality of life and happiness. Additionally, one could measure objective health outcomes such as blood pressure and body mass index. Longitudinal research designs are required, where outcome measures are assessed before, right after and a long time after the health tourism programme, to evaluate the possibility of short-term versus long-term and permanent change. We also need to find a way to assess if the *tourism* component makes any difference to wellness outcomes.

Finally, we acknowledge that our categorisation of the 'pure' medical and wellness tourism providers does in some cases exaggerate the differences between them. For instance, some lifestyle resorts such as the *Chiva-Som* in Thailand employ mainstream health professionals like physicians and nurses and also include medical diagnostics to assess the health status of clients. On the other hand, some corporate investor-owned hospitals have started to incorporate 'wellness centres', like the *Bumrungrad International Hospital* in Thailand with its *VitalLife Wellness Center*. These examples show that the boundaries between medical and wellness tourism providers are far from rigid.

Conclusion

- The typology of medical and wellness tourism providers in this chapter is designed to assist understanding of this emerging and volatile industry.
- It should be noted however that the boundaries between the different types of provider are often blurred and changing in line with societal changes and developments.
- There are important ethical issues to be confronted in the supply of health tourism, including access to treatment and supplying services that conflict with the principles of the wellness paradigm.
- Future research needs to explore these dilemmas in greater depth, as well as the trend towards collaboration in the delivery of health care, and the structure and efficacy of different forms of medical and wellness tourism provided across the globe.

This chapter illustrates the changing and blurred nature of medical and wellness tourism providers. Many of the structural developments of health tourism are recent, especially the emergence of the five medical wellness tourism provider types. We have discussed broader changes in consumer culture and the provision of health care that have influenced the growth of health tourism in general and the continuous diversification and blurring of 'pure' medical and wellness tourism providers in particular. While our current typology of health tourism is hopefully beneficial to understand the current supply structure of health tourism, its categories are neither neatly discrete nor permanent. Health tourism supply will continue to shift and change according to broader cultural, political and economic developments. As long as there is no standardised way to differentiate effective from non-effective treatments, most health tourism provisions which are largely based on elective, partially non-necessary medical treatments, as well as different mixes of CAM, lifestyle-based interventions and integrative approaches, will continue to be unsupported by public health schemes. Global privatised health care, which includes the largely privatised health tourism sector, will be accessible only to those who can afford to select their preferences from an increasingly diversified potpourri of health-care options.

References

Ackerman, S. L. (2010) 'Plastic paradise: Transforming bodies and selves in Costa Rica's cosmetic surgery tourism industry' *Medical Anthropology* 29, 4, 403–23.

Askegaard, S., Gertsen, M. C. and Langer, R. (2002) 'The body consumed: Reflexivity and cosmetic surgery' *Psychology and Marketing* 19, 10, 793–812.

Baer, H. A. (2002) 'The growing interest of biomedicine in complementary and alternative medicine: A critical perspective' *Medical Anthropology Quarterly* 16, 4, 403–5.

Baer, H. A. (2008) 'The emergence of integrative medicine in Australia: The growing interest of biomedicine and nursing in complementary medicine in a Southern developed society' *Medical Anthropology Quarterly* 22, 1, 52–66.

Baer, H. A. and Coulter, I. (2009) 'Introduction – Taking stock of integrative medicine: Broadening biomedicine or co-option of complementary and alternative medicine?' *Health Sociology Review* 17, 4, 331–41.

Bodeker, G. and Kronenberg, F. (2002) 'A public health agenda for traditional, complementary, and alternative medicine' *American Journal of Public Health* 92, 10, 1582–91.

Bookman, M. Z. and Bookman, K. R. (2007) *Medical Tourism in Developing Countries* (New York: Palgrave Macmillan).

Carmona, R. H. (2008) *Prevention and Global Health: The Vital Role of Spas.* Paper presented at the Global Spa Summit.

Clinique La Prairie (2011) 'Philosophy' http://www.laprairie.ch/en/clinique-la-prairie/philosophy, date accessed 12 November 2011.

Cohen, E. (2008a) 'Medical Tourism in Thailand' in E. Cohen (ed.) *Explorations in Thai Tourism* (Bingley, UK: Emerald).

Cohen, M. (2008b) 'Spas, Wellness and Human Evolution' in M. Cohen and G. Bodeker (eds) *Understanding the Global Spa Industry: Spa Management* (Oxford: Butterworth-Heinemann).

Connell, J. (2006) 'Medical tourism: Sea, sun, sand and... surgery' *Tourism Management* 27, 6, 1093–100.

Dalen, J. E. (1999) 'Is integrative medicine the medicine of the future? A debate between Arnold S. Relman, MD, and Andrew Weil, MD' *Archives of Internal Medicine* 159, 18, 2122–26.

Davis-Floyd, R. (2001) 'The technocratic, humanistic, and holistic paradigms of childbirth' *International Journal of Gynaecology and Obstetrics* 75, 1, S5–S23.

Dobos, G. (2009) 'Integrative medicine–medicine of the future or "old wine in new skins"?' *European Journal of Integrative Medicine* 1, 3, 109–15.

Dunn, H. L. (1959) 'What high-level wellness means' *Canadian Journal of Public Health* 50, 11, 447–57.

Eisenberg, D. M., Davis, R. B., Ettner, S. L., Appel, S., Wilkey, S., Van Rompay, M. and Kessler, R. C. (1998) 'Trends in alternative medicine use in the United States, 1990–1997: Results of a follow-up national survey' *The Journal of the American Medical Association* 280, 1569–75.

Ellis, S. (2008) 'Trends in the Global Spa Industry' in M. Cohen and G. Bodeker (eds) *Understanding the Global Spa Industry: Spa Management* (Oxford: Butterworth-Heinemann).

Ernst, E. (2011) 'Integrated medicine: Smuggling alternative practices into rational medicine?' *Focus on Alternative and Complementary Therapies* 16, 1, 1–2.

Featherstone, M. (1991) 'The Body in Consumer Culture' in M. Featherstone, M. Hepworth and Turner, B. S. (eds) *The Body: Social Process and Cultural Theory* (London: Sage).

Gaudet, T. W. and Faass, N. (2001) 'Developing an Integrative Medicine Program: The University of Arizona Experience' in N. Faass (ed.) *Integrating Complementary Medicine into Health Systems* (Aspen: Gaithersburg).

Gimlin, D. (2000) 'Cosmetic surgery: Beauty as commodity' *Qualitative Sociology* 23, 1, 77–98.

Goldstein, M. S. (2000) 'The Culture of Fitness and the Growth of CAM' in M. Kelner, B. Wellman, B. Pescosolido and M. Saks (eds) *Complementary and Alternative Medicine: Challenge and Change* (London: Routledge).

Hattie, J. A., Myers, J. E. and Sweeney, T. J. (2004) 'A factor structure of wellness: Theory, assessment, analysis and practice' *Journal of Counseling and Development* 82, 3, 354–64.

Hollenberg, D. (2007) 'How do private CAM therapies affect integrative health care settings in a publicly funded health care system?' *Journal of Complementary and Integrative Medicine* 4, 1, 1–16.

Kaptchuk, T. J. and Miller, F. G. (2005) 'Viewpoint: What is the best and most ethical model for the relationship between mainstream and alternative medicine: Opposition, integration, or pluralism?' *Academic Medicine* 80, 3, 286–90.

Kelly, C. (2010) 'Analysing wellness tourism provision: A retreat operators' study' *Journal of Hospitality and Tourism Management* 17, 1, 108–16.

Kelner, M. and Wellman, B. (1997) 'Health care and consumer choice: Medical and alternative therapies' *Social Science and Medicine,* 45, 2, 203–12.

Kelner, M. M. and Wellman, B. (2003) 'Complementary and alternative medicine: How do we know if it works?' *Healthcare Papers: New Models for the New Healthcare* 3, 5, 10–28.

Lempa, H. (2008) 'The spa: Emotional economy and social classes in nineteenth-century Pyrmont' *Central European History* 35, 1, 37–73.

McGinley, S. (2011) 'Princess Haya to Steer Revamp of Dubai Healthcare City', *Arabian Business,* http://www.arabianbusiness.com/princess-haya-steer-revamp-of-dubai-healthcare-city-422610.html date accessed 12 November 2011.

Menzel, D. (2011) 'Ausgezeichnet kuren und sich wohlfühlen' *Nordkurier* 56.

Müller, H. and Lanz Kaufmann, E. (2001) 'Wellness tourism: Market analysis of a special health tourist segment and implications for the hotel industry' *Journal of Vacation Marketing* 7, 1, 5–17.

Nahrstedt, W. (2004) 'Wellness: A New Perspective for Leisure Centers, Health Tourism, and Spas in Europe on the Global Health Market' in K. Weiermair and C. Mathies (eds) *The Tourism and Leisure Industry: Shaping the Future* (New York: Haworth Hospitality Press).

O'Connor, B. (2000) 'Conception of the Body in Complementary and Alternative Medicine' in M. Kelner, B. Wellman, B. Pescosolido and M. Saks (eds) *Complementary and Alternative Medicine: Challenge and Change* (London: Routledge).

Puczkó, L. and Bacharov, M. (2006) 'Spa, bath, thermae: What's behind the labels' *Tourism Recreation Research* 31, 1, 83–91.

Ryan, R. M. and Deci, E. L. (2001) 'On happiness and human potentials: A review of research on hedonic and eudaimonic well-being' *Annual Review of Psychology* 52, 141–66.

Saks, M. (2000) 'Professionalization, Politics and CAM' in M. Kelner, B. Wellman, B. Pescosolido and M. Saks (eds) *Complementary and Alternative Medicine: Challenge and Change* (London: Routledge).

Saks, M. (2003) *Orthodox and Alternative Medicine* (London: Continuum).

Shaw, I. and Aldridge, A. (2003) 'Consumerism, health and social order' *Social Policy and Society* 2, 1, 35–43.

Sheldon, P. J. and Bushell, R. (2009) 'Introduction to Wellness and Tourism' in R. Bushell and P. J. Sheldon (eds) *Wellness and tourism: Mind, body, spirit, place* (New York: Cognizant).

Smith, M. and L. Puczkó (2008) *Health and Wellness Tourism* (Oxford: Butterworth-Heinemann).

Snyderman, R. and Weil, A. (2002) 'Integrative medicine: Bringing medicine back to its roots' *Archives of Internal Medicine* 162, 4, 395–7.

TRAM (2006) *Medical Tourism: A Global Analysis* (Arnhem: Association for Tourism and Leisure Education – ATLAS).

Travis, J. W. and Ryan, R. S. (1988) *The Wellness Workbook*, 2nd edn (Berkeley, CA: Ten Speed Press).

Verschuren, F. (2004) *Spa Health and Wellness Tourism: A New Product Portfolio at the Canadian Tourism Commission* (Ottawa: Canadian Tourism Commission).

Voigt, C., Brown, G. and Howat, G. (2011) 'Wellness tourists: In search for transformation' *Tourism Review* 66, 1/2, 16–30.

Voigt, C., Laing, J., Wray, M., Brown, G., Howat, G., Weiler, B. and Trembath, R. (eds) (2010) *Wellness and Medical Tourism in Australia: Supply, Demand and Opportunities* (Gold Coast: CRC for Sustainable Tourism).

Weisz, G. (2001) 'Spas, mineral waters, and hydrological science in twentieth-century France' *Isis* 92, 3, 451–83.

Williamson, C. (1999) 'Reflections on health care consumerism: Insights from feminism' *Health Expectations* 2, 3, 150–8.

4

Caring for Non-residents in Barbados: Examining the Implications of Inbound Transnational Medical Care for Public and Private Health Care

Jeremy Snyder, Valorie A. Crooks, Leigh Turner, Rory Johnston, Henry Fraser, Laura Kadowaki, Mary Choi and Krystyna Adams

Introductory summary

- Barbados is a tourism-dependent island state whose income is very sensitive to perturbations in the global economy and is thus seeking ways to diversify its service exports.
- Barbados has a two-tiered health system, with a publicly funded health system operating alongside numerous private clinics.
- Private provision of health services in Barbados has grown between 2000 and 2010.
- Barbados has an established history of providing health care to ill vacationers, to other Caribbean residents and, more recently, to a small but growing number of medical tourists seeking fertility care.
- The Bajan government is aggressively promoting the development of their medical tourism industry, including the facilitation of the development of a mid-size private hospital catering primarily to medical tourists.
- The development of medical tourism services in Barbados carries with it not only the potential for some economic gains but also significant risks to health equity if mismanaged or left unregulated.

Introduction

Barbados, the most easterly island in the Caribbean, is a small nation totalling 430 square kilometres with a population of just under 300,000 people. Its population is rapidly ageing, due to a falling fertility rate and rising life expectancy. The over-65 population is 9.8 per cent, second highest in the Caribbean, compared to 4.8 per cent in Guyana and 6.3 per cent in the Bahamas, and close to 13.1 per cent for the US (PAHO, 2007). Much like other island economies, it is essential that Barbados brings revenue and other resources into the country in order to continue economic development and sustain public services such as health care. In 2002, the country's national government developed the 'Strategic Plan for Health', which aims to reform the health-care system in response to the changing health needs of the population (PAHO, 2007). Government officials are aware of the financial resources required to implement the reforms put forth in this initiative and other health-care-related proposals.

The Bajan economy is heavily dependent on tourism, and many jobs are related to this industry. In 2009, as a result of the global recession, 519,000 stay-over tourists visited the island, a number which decreased by 8.7 per cent from 2008, demonstrating Barbados' vulnerability to global economic downturns (PAHO, 2007; BBC, 2011). The dominance of the tourism sector in Barbados' economy means that tourism and health care are intertwined, in that public health-care resources are supported by tourism revenues in the form of tax revenues and employment. In this chapter we examine a different dimension of this link through exploring the ways in which the country is involved in inbound transnational medical care, or, in other words, providing medical care for non-locals.

Although Barbados is a small country both in terms of physical size and population, it maintains important ties with other nations. These economic, social and familial bonds impact both inbound flows of non-locals and the country's involvement in provision of transnational medical care. Barbados is a former British colony that gained independence in 1966 (BBC, 2011). Approximately 4500 British citizens currently reside on the island, and Barbados receives a large portion of imports such as food and beverage and manufactured articles from the UK, the US and other Caribbean countries, while also exporting materials such as sugar to the UK (Foreign and Commonwealth Office, 2010). As we discuss later in this chapter, Barbados also receives significant numbers of long- and short-stay tourists from the UK, its major source of stay-over visitors.

Barbados also maintains strong relations with regional countries through its involvement in the Caribbean Community (CARICOM), an organisation that benefits member countries through regional cooperation, in particular trade (Foreign and Commonwealth Office, 2010; CARICOM, 2011). While this trade has numerous effects upon Barbados' economy, in this chapter we pay particular attention to how Barbados' regional networks throughout the Caribbean – through traditional migration patterns, education and training programmes, regional commercial activities, regional tourism and ties developed by CARICOM, and other networking bodies – have resulted in the development of *regional* transnational medical care flows into the country.

In the remainder of this chapter we examine Barbados' involvement in the provision of inbound transnational medical care. Our main purpose is to articulate the different types of inbound transnational medical care offered within the country and describe the significance of such care provision for Barbados' public and private health-care sectors. To provide adequate context, we begin with a brief overview of Barbados' health system. We then review findings from stakeholder interviews conducted in 2011 to provide a detailed account of the three types of inbound transnational medical care that currently operate on the island: (1) care for ill vacationers, (2) care for medical tourists and (3) regional medical care. Each of these types of care involves provision of treatment to a different group of non-local individuals. After discussing the implications of these types of transnational medical care for Barbados' public and private health-care sectors, we conclude by identifying future research directions regarding inbound transnational medical care within Barbados specifically and the Caribbean more generally.

Understanding Barbados' health system

By the time Barbados achieved independence in 1966, much of the infrastructure for the delivery of its social programmes was already in place due to the country's long history of commitment to social development (ECLAC, 2001; Rodney and Copeland, 2009). Within the health sector, major developments in the establishment of Barbados' health system include the construction of public health centres in the 1950s for the provision of public health services. Limited free primary care had been provided for many years by a cadre of 12 Parochial Medical Officers (one doctor in each parish except for St. Michael, where there were two) and parish infirmaries for elderly and indigent patients. In the 1960s a more comprehensive primary health-care system was initiated, to complement the opening of the newly built 550-bed Queen Elizabeth Hospital (QEH) in 1964, expanded training facilities for nurses (ECLAC, 2001) and started training of medical students of the University of the West Indies, in partnership with the QEH. In 1969, passage of the Health Services Act established the legal framework governing the delivery of comprehensive, public health care in Barbados. The Ministry of Health is responsible for the administration of the Act and is entrusted with ensuring the health of the population (PAHO, 2008).

Current health-care facilities

Health-care services in Barbados are available through a network of primary, secondary and tertiary institutions in both public and private sectors (PAHO, 2008). In the public sector, a network of eight polyclinics and four satellite clinics located throughout the island provides a wide range of primary care and public health services (PAHO, 2001). The government-run Queen Elizabeth Hospital located in the capital city of Bridgetown is the main public health facility for acute, secondary, tertiary and emergency health-care services, and it also functions as a teaching and research hospital for the University of West Indies (PAHO, 2005). In 2011 the government announced plans to replace the 47-year-old hospital with a new BBD$800 million plus, 600-bed facility through a joint public, private and international financing (Lashley, 2011). Other public facilities include a network of four district hospitals for geriatric care, a main geriatric institution, a mental health hospital, two small rehabilitation institutions, an AIDS hostel and a HIV/AIDS Diagnostic and Treatment centre, Children's Development Centre, National Nutrition Centre, Family Planning Unit and renal dialysis facilities (PAHO, 2001, 2008).

Health system challenges

While Barbados has made great achievements over the past century in both the delivery of its health care system and the health of its population, it nevertheless faces many significant challenges. Barbados' growing elderly population will increase demand for treatment of chronic, non-communicable diseases as well as provision of both rehabilitative services and specialised geriatric care services (ECLAC, 2001). Health ministry officials anticipate having to address increased burdens associated with managing obesity, diabetes and assorted chronic, non-communicable diseases (PAHO, 2008). These problems serve to exacerbate differential access to care, differences in health status within the population and inequity within the health-care system. Lastly, the health sector has a major shortage of some categories of health workers, particularly at the primary health-care level (PAHO, 2001). Government officials hope to address loss of health-care professionals both by reducing emigration of health-care providers and increasing the number of health-care workers trained at regional community colleges and universities (PAHO, 2008).

Understanding Barbados' inbound transnational medical care

In this section we examine Barbados' three main types of *inbound* transnational medical care: (1) care for ill vacationers, (2) care for medical tourists and (3) regional medical care. The review provided here is informed by a review of relevant scholarship and policy documents, interviews conducted with stakeholders in Barbados' public and private health service sectors in 2011, and site visits to numerous health-care facilities in Barbados. In total, 19 semi-structured interviews were conducted in person or by phone. These participants were drawn from health care, tourism, government and medical tourism-affiliated sectors in Barbados, Canada and the US. Both public and private organisations were represented. These interviews examined the nature of Barbados' health system, including its structure and challenges, and the country's growing medical tourism industry. As our interviews focused on inflows of international patients into Barbados, the types of transnational medical care that the stakeholders most frequently discussed dealt with inbound flows. As such, this serves as our focus here.

Care for ill vacationers

Most health services in Barbados are accessible to foreign visitors who fall ill during the course of their vacation. There are two main types of visitors: stopover tourists who typically arrive by air and stay for at least one night and up to a year, and cruise ship passengers who generally stay less than a day (Jessen and Vignoles, 2004). In 1990, the last time such data was collected, foreign visitors represented 2.1 per cent of all admissions and 1.6 per cent of all accident and emergency visits to the Queen Elizabeth Hospital, at a cost of approximately BDS$1.1 million per year ($550,000 USD) or 2 per cent of all hospital costs (Walters, Fraser and Alleyne, 1993; Gonzales, Brenzel and Sancho, 2001). In the past fee collection was often inefficient, especially for emergencies, but efforts are being made to improve fee recovery.

Several health facilities treat visitors to Barbados. In particular, cruise ship passengers are becoming an important client base for health providers who hope to benefit from Barbados' reputation as an established medical destination among cruise ship lines operating in the Caribbean (Gonzales, Brenzel and Sancho, 2001; Jessen and Vignoles, 2004). BayView Hospital notably has an arrangement with cruise lines that visit Barbados to provide needed medical attention to their passengers (Paffhausen et al., 2010). Barbados is also home to Island Dialysis, one of several facilities in the Caribbean operated by Canadian-based Atlantic Healthcare Group, Inc. (Gonzales, Brenzel and Sancho, 2001). In addition to providing dialysis to local residents, the clinic markets its services to vacationers planning to holiday in Barbados.

Over the past few decades, the demographics of foreign visitors to Barbados have shifted towards older populations. In 2006, persons over the age of 50 represented 28 per cent of visitors and the largest age group of all stopover visitors to the island (Ministry of Tourism, 2009). The total number of such arrivals has been steadily increasing from 103,327 since 1989 to 159,556 in 2006 (Gonzales, Brenzel and Sancho, 2001; Ministry of Tourism, 2009). In part as a response to this trend, specialty health services targeting seniors for long-term resident care, such as nursing homes, have been identified as a possible area of development and economic opportunity (Gonzales, Brenzel and Sancho, 2001).

Care for medical tourists

Medical tourists are differentiated from ill vacationers or visitors on the basis of intentionality (Crooks et al., 2010). Vacationers who become ill while in Barbados do not travel to the country intending to access medical care, while medical tourists typically do, though aggressive marketing to visitors and/or the availability of inexpensive medical care can sometimes prompt vacationers to intentionally or purposefully access non-emergency medical care while abroad. Those who do so typically obtain cosmetic procedures and/or dental care as opposed to major surgical procedures that require extensive planning, testing and post-operative care. Medical sites that treat international patients in Barbados now include the Barbados Fertility Centre (BFC), Island Dialysis, the Sparman Clinic, the BayView Hospital and the Sandy Crest Medical Centre. A new medical tourism site, American World Clinics (AWC), is presently in development. Currently the BFC, which opened in 2002, is considered to be the frontrunner in Barbados' medical tourism industry. It is an in vitro fertilisation (IVF) treatment centre and the first full fertility unit located in the Caribbean. It is also the only medical tourism destination in Barbados that currently has Joint Commission International accreditation. BayView Hospital and the Sandy Crest Medical Centre are viewed as offering private medical care without marketing health care to patients from outside the Caribbean; therefore, their participation in the medical tourism sector is limited, although it is worth noting that the predecessor of the BayView Hospital, the smaller Diagnostic Clinic, was a significant medical tourism referral centre for the Eastern Caribbean in the past. Island Dialysis is unique in that its services mainly help to enable tourists to travel to Barbados, rather than to attract them to the island.

The Government of Barbados is interested in expanding the scope of the medical tourism sector in Barbados. As one interviewee commented, 'there's support for the expansion of the medical tourism industry, and that's quite obvious from the amount of work Invest Barbados, who is a government owned agency, is putting into the whole marketing of medical tourism from Barbados'. AWC's hospital facility, which is scheduled to open in 2013, will build upon the precedent established by BFC. According to company marketing material, the facility will target international patients and is expected to contain 12 operating rooms, 50 inpatient beds and 20 outpatient beds. AWC will mainly offer outpatient and short-stay procedures, including cosmetic dentistry, plastic surgery, ophthalmic surgery and orthopaedic surgery. Physicians staffing the clinic will primarily be from the US, with some local physicians and surgeons, while local Bajans will be hired for the non-physician staff. Local trade specialists estimate that development of the hospital facility will lead to the creation of 230 jobs, as well as the generation of approximately BDS$50 million plus per year in revenue from international patients and their travel companions for Barbados (Invest Barbados, 2011). Patients from such countries as the US, the UK and Canada will be targeted for this facility.

The AWC development is a sign of the Barbados government's increasing commitment to developing the medical tourism industry. As early as

1994, the island's national government began showing interest in medical tourism when, in partnership with the World Bank, it commissioned a report assessing the potential of health tourism (Gonzales, Brenzel, and Sancho, 2001). In recent years, there has been a renewed interest in medical tourism, spurred, at least in part, by the success of BFC. For example, in 2008 Barbados hosted the Exporting Services to Canada Seminar, and Health and Wellness Tourism Strategy Session in order to identify ways Barbados can export medical and health services to inbound North Americans (Caribbean Export Development Agency, 2011). Barbados has also created a Health and Wellness Task Force. In 2009, this organisation provided recommendations concerning ways to facilitate the development of the medical tourism industry. These recommendations included creating a medical tourism development plan, passage of new legislation, creating a wellness council and establishing appropriate coordinating bodies (Gill, 2010). As a result of these initiatives, stakeholders in the public and private health-care sectors regard the national government as a keen supporter of the country's emerging medical tourism industry.

The Barbados International Business Promotion Corporation, a quasi-governmental organisation that is better known by the name it trades under, Invest Barbados, is funded by the national government for the purpose of promoting international business. Invest Barbados is a key player in the development and promotion of the medial tourism industry in Barbados. Invest Barbados currently provides support to the medical tourism facilities already in operation in Barbados. Its employees also work with investors to create new opportunities in the industry.

Regional medical care

A strong foundation for medical cooperation has been built by Caribbean nations due to the establishment of CARICOM in 1973, with a very modestly staffed 'health desk'. This organisation promotes regional cooperation within the Caribbean, largely of a policy and regional planning nature. While some facilitation is offered to international (Caribbean region) patients, it is primarily a government-facilitated, cross-border care system. As one of our interviewees commented, 'there are some programs at the Queen Elizabeth Hospital...I wouldn't qualify them as medical tourism per se because the patients are coming from other countries but it's really not under the aegis of a tourism type of activity, it's more of a... traditional referral pattern'.

Barbados plays a significant role in providing medical care for residents of other Caribbean nations. It offers inbound transnational medical care for these individuals, sometimes through cross-border care arrangements. We use the term 'cross-border care arrangements' here to refer to medical care provided to a citizen of one country in another country that is financed by the public health-care system from his/her home country. Within the

Caribbean, Barbados is regarded as a favoured destination for regional patients, particularly for those from smaller islands lacking advanced diagnostic and treatment facilities and the capacity to offer treatment to high-risk patients. Regional medical travellers also arrange private medical care in Barbados. BFC, the Sparman Clinic, Island Dialysis and BayView Hospital all attract private regional patients; however, according to our interviewees, the public Queen Elizabeth Hospital is the primary health-care destination for regional patients.

The Queen Elizabeth Hospital serves as the main referral hospital for the entire Eastern Caribbean (Gonzales, Brenzel, and Sancho, 2001). Governments of other Caribbean nations refer patients via established cross-border care agreements to the Queen Elizabeth Hospital through Barbados' Ministry of Health Chief Medical Officer. As one stakeholder noted, 'All consultants have admitting privileges for private patients and what we find is that because we have a number of specialities here that some of the smaller islands find it difficult to have, patients are oftentimes referred from many of the OECS countries, to our consultants for care.' Consultants at the Queen Elizabeth Hospital also have the ability to admit private patients such as ill vacationers not covered by the island's public system and other non-locals purposely seeking care in Barbados outside existing cross-border care arrangements. In these cases patients pay for their care out of pocket according to a fee schedule partly set by the Ministry of Health. There are also facilities intended to service not only Barbados but also the region. Examples of such hospital units include the orthopaedics and radiotherapy departments at the Queen Elizabeth Hospital and the regional optometric centre.

Implications of transnational medical care for public and private health care in Barbados

Transnational medical care in Barbados has many implications, both for the provision of medical care within Barbados and for the country's economic development. Within the publicly funded health-care system, transnational medical care has the potential for both positive and negative effects. The provision of transnational care could generate increased funding for the public system if tourists, medical tourists and regional patients provide payments that are captured by the public system. This potential is weakest from medical tourism as these patients typically target private health providers. When visitors are admitted to the QEH, they may be admitted to a public ward or intensive care unit if very ill and requiring extended emergency care. In these cases, fees are paid directly to the hospital. When admitted to private units, patients pay fees both to the hospital and to the individual consultant as private fees.

Barbados could address this issue by developing a national medical tourism strategy with the potential to generate greater benefits for the public system (Bookman and Bookman, 2007). For example, payment of corporate taxes generates revenues for the Barbados government; these tax dollars can then be used to support publicly funded health care. In addition, the Barbados government could impose a tax or surcharge on all procedures or other health-care interventions that are provided to medical tourists by private, for-profit health-care facilities. These tax dollars could be used to cross-subsidise the publicly funded health-care system. While there are several practical obstacles to such a taxation scheme, in theory, and perhaps in practice, if adequate monitoring and financial mechanisms were established, such an approach could benefit Barbados' domestic publicly funded health system.

Acknowledging that provision of transnational medical care could be used to promote economic and social development in Barbados, it is important to note that providing care for international patients could, at least in theory, have the deleterious effect of drawing attention and resources away from the needs of the local population and the public health care system. If medical tourists have medical needs different from those of the local population, then local patients might in time find it more difficult to access other forms of care, including more cost-effective preventive medicine and care for chronic conditions. These concerns are primarily generated by inflows of vacationers and medical tourists with medical emergencies; regional patients who access care in Barbados through planned, cross-border care arrangements likely can be more easily managed in ways that support rather than hinder the public health system, but in all cases international patients have the potential to problematically re-focus health priorities within Barbados to the detriment of the local population if there should be a rapid influx without health planners planning for or recognising it. Thus, anticipating harms and planning to mitigate negative effects while maximising collective benefits are crucial. Absent adequate safeguards, expansion of medical tourism within Barbados could harm health equity, publicly funded health care facilities, health human resources and access to public health care. With proper planning, the Barbados government might be able to avoid such harms while benefiting the domestic health-care sector, the larger economy and non-local citizens seeking access to health care. Indeed planning facilities for medical tourism could provide benefits to the local population.

In addition to the public health care system, Barbados has a flourishing private, for-profit health care system that provides health services to both local residents and international patients. As Barbados has a small population, one possible advantage of opening the private health care sector to more international patients is that the range of specialised services offered is likely to be greater than what could be supported exclusively by local residents. Expansion of the private health care sector might enable Bajans to access a greater range of medical services without travelling abroad for care than would be otherwise possible. Sufficient patient volume would be needed for there to be adequate financial justification for increased specialisation of health services.

While increased specialisation of health care has potential to benefit some Bajans, it is important to acknowledge that unless Barbados' Ministry of Health takes deliberate steps to promote health equity and provide government-funded access to such specialised health services, these enhanced options would benefit only those individuals with the financial means or insurance coverage to pay for these services. In short, there is no reason to assume that expansion of the private health sector and increased specialisation of health services will benefit all Bajans. Moreover, if the private medical sector in Barbados focuses on the needs of international patients and specialises in certain forms of care, then this enhanced service access will be narrower than a broad increase in services targeted at meeting the full range of needs of the local population. For example, if medical facilities targeting international patients focus on specific areas such as fertility treatment and elective surgeries, these services will be of use to the members of the local population who need and can afford access to them. The services offered will be different and, potentially, more limited, however, than if they were being targeted solely at the local population. That point noted, even a narrowly focused expansion of local private care can benefit some local residents when compared to the offerings available without the support of international patients. Also, improved facilities and specialist care designed to cater for medical tourism would obviate the need for many patients, usually those with health insurance, to travel overseas for investigation or care, thus saving considerable foreign exchange, with direct benefit to the economy. Furthermore, the government could more readily afford to utilise such facilities for public patients than to authorise overseas care.

The inflow of international patients also has implications for health human resources. Historically, for numerous reasons Barbados has had difficulty retaining trained health workers. The draw of higher-paying jobs regionally and internationally pulls many skilled health-care professionals away from Barbados. Because English is the national language of Barbados, these workers are especially attractive to employers in the US, the UK and Canada. A potential positive impact of increased inflow of international patients is that they can help support health sector jobs within Barbados. Better opportunities for Bajans trained in health professions might slow or reverse the emigration of trained health-care providers. With greater demand for medical services can come better pay, working conditions and training opportunities, all of which will reduce the appeal of migration.

However, if vacationers and medical tourists are principally accessing private care, expansion of the private sector might undermine public care within Barbados by drawing health human resources to higher-paying private positions. Thus the problem of supplying adequate health human resources in the public sector in Barbados may simply be shifted if health-care professionals in the public sector relocate to the private sector rather than migrate to other countries. For Bajans unable to afford access to the private sector, these workers will be just as inaccessible to them as they would be were they located in another country. Given this possibility, it is important that the Barbados government take realistic steps to reduce flows of workers from the public to private sector as flows of international patients increase or that it develop ways for private sector human resources to support the public sector. Though at present there is no indication of mass movement of health-care professionals from Barbados' publicly funded health system to the private, for-profit health sector, researchers studying other medical tourism destinations report that expansion of private health care sector to treat international patients can result in movement of health care personnel from publicly funded medical facilities to private, for-profit hospitals and clinics. In our interviews with medical tourism stakeholders in Barbados, this concern was clearly articulated and it was discussed that a memorandum of understanding between the government and AWC included provisions for managing impacts on health human resources. The details of this agreement have not been publicised at this time.

In addition to its implications for the public and private medical sectors in Barbados, inflows of international patients have important implications for other sectors, including tourism and business development. This is the case particularly for medical tourists who are encouraged to couple their medical treatment with time enjoying the climate and culture of the country, and in so doing contribute to a range of businesses in the hospitality industry. Barbados' temperate climate combined with its well-established hospitality sector might lead to the development of particular surgical offerings that are conducive to recovery or preparatory periods for patients, as has been the case with BFC. The economic impact of such forms of care differs from other types of treatment where patients remain in Barbados for a brief period and broader economic benefits associated with their visits are limited. While the economic impacts of ill vacationers and regional patients may be more limited as they are less likely to plan extended stays within Barbados, even in these cases family members may accompany these patients, stay in hotels, use local restaurants, rent vehicles or taxis and otherwise have an impact upon the local economy. Moreover, extended recovery times will support local non-medical businesses.

There are a number of factors about Barbados that make it an appealing medical tourism destination, including the climate, proximity to Canada and the US, skilled health practitioners, good infrastructure and a strong tourism sector (Jessen and Vignoles, 2004). Moreover, Barbados benefits from the fact that English is spoken there with little dialect use among professionals, as opposed to Middle East and Far Eastern countries. It is conceivable that stiff competition in the medical tourism sector will mean that Barbados will not be able to capitalise on an influx of international patients and subsequent economic benefits. Some researchers are sceptical that Barbados will ever be able to compete with leading medical tourism destinations in Asia and Latin America, especially on cost (Chambers and McIntosh, 2008). If the medical tourism sector in Barbados severely underperforms expectations, then public resources devoted to encouraging this market will have been wasted.

The provision of care to international patients and, especially, the expansion of the private, for-profit health care sector in Barbados prompts concerns about health equity for the local population. Several stakeholders mentioned this concern in our interviews, though we were also informed that the Barbados government is committed to ensuring that increased medical tourism will not affect patient access. As we have observed, an influx of international patients might enhance access to medical services and provide economic benefits throughout the community. However, widening differential access to medical care is a concern in itself even if the net effect of international patients entering the country is positive. These effects will need to be closely monitored, managed and mitigated if health equity is to be supported by the provision of care to international patients.

Future research directions

It is possible that Barbados will not see a significant increase in inbound transnational medical care. The global marketplace for health services is extremely competitive and health-care facilities in many countries compete for international patients. However, Barbados' relative proximity to the US, Canada and other Caribbean nations, as well as its longstanding ties to the UK, suggests that it might have success in attracting medical tourists from these countries. In addition, Barbados has historically attracted numerous regional medical travellers from other Caribbean countries and with appropriate planning the QEH is best placed to capitalise on the Caribbean referrals and to greatly expand in this area. Public and private sector initiatives to promote Barbados as a medical tourism destination, the country's longstanding appeal as a tourist destination as well as its developed hotel and resort industry suggest that Barbados has some advantages as it tries to increase its standing as a destination for transnational medical care.

Recognising that government and private sector efforts to attract international patients to Barbados could lead to increased inflow of international patients, we identify numerous topics as warranting further research. Of the numerous hospitals and clinics in Barbados, BFC and AWC will be worth studying as they try to attract non-local residents to Barbados. BFC has already been successful in establishing itself as a clinic capable of attracting an international clientele. AWC remains at the planning stage, but its scale suggests that it could play a significant role in expanding the place of transnational medical care in Barbados. In addition to studying particular health-care facilities, we see need for careful quantitative, qualitative and ethical analysis of the effects of increased inflow of international patients. With regard to quantitative data, both publicly funded health-care facilities and private, for-profit health-care institutions should carefully document how many international patients they treat, the types of medical tests and treatments they receive, the home countries from which they travel, financial losses or gains associated with treating them, and both morbidity and mortality associated with providing care to international patients. Such data should be provided to Barbados' Ministry of Health and used to assess the effects of increased medical travel, including the spin-off tourist effects of relatives and friends. In addition, health ministry officials, local Bajan researchers and international researchers can all contribute to studying the health system, health human resources and health equity effects of increased transnational medical care within Barbados. If Barbados succeeds in attracting significant numbers of international patients, the for-profit, private health-care sector could expand relative to the publicly funded health system. Taxation of procedures provided to international patients could generate revenues that can then be used to cross-subsidise care of local residents. However, it is also possible that the public health system will experience harms rather than benefits as a result of increased transnational care.

We regard transformation of Barbados' health-care sector, as well as the possible expansion of Barbados' private health-care sector, as a topic meriting careful study over the next five to ten years. The effects of medical travel upon health human resources are another subject warranting additional research. Increased transnational medical care could help Barbados retain health-care professionals and, importantly, expand provision of numerous specialised health services. However, it is also possible that increased transnational medical care will lead to intranational 'brain drain' of health-care providers from Barbados' publicly funded health-care facilities to private, for-profit hospitals and clinics. We regard the study of health human resources' effects of increased medical travel as a subject deserving serious scholarly analysis. Next, increased transnational medical care could promote domestic health equity, if revenue generated from treating international patients is used to cross-subsidise care of local citizens. The health equity effects of increased transnational medical care in Barbados deserve thorough analysis as both government officials and local entrepreneurs try to increase Barbados' standing as a destination for international patients. Finally, we suggest the importance of studying Barbados' emerging medical tourism industry in a comparative perspective. Both within the Caribbean and the larger global health services marketplace, countries increasingly are seeking to use medical tourism as a form of economic and social development, attracting international patients and fusing provision of health care with domestic hospitality industry.

We anticipate that there is much to gain from using numerous research methods to compare and contrast developments in Barbados with health systems, health human resources, health equity and health policy changes in other countries striving to attract international patients. Studying such transformations requires methods, methodologies and theories from many different scholarly disciplines, and as we continue to study Barbados as a destination for transnational medical care we intend to draw from such fields as geography, bioethics, philosophy and health sciences.

Concluding summary

Barbados hopes to emerge as a popular medical tourism destination, as demonstrated by government efforts to develop its industry, although its potential success or failure is far from clear. Increasing medical tourism flows to Barbados may ease or alter existing health and human resources shortages in the country depending on how this industry is managed. If equitably managed, increasing medical tourism could improve the quality of domestic health services by supporting medical specialisation, helping health worker retention as well as cross-subsidising care for domestic users by foreign patients. If poorly regulated, increasing medical tourism in Barbados could exacerbate health inequities by facilitating the growth of the private health-care sector. Assessing the impacts of medical tourism in Barbados will take time and would require input from a variety of disciplines and stakeholders

Acknowledgements

This research was funded by a Planning Grant awarded by the Canadian Institutes of Health Research.

References

BBC (2011) *Barbados Country Profile* http://news.bbc.co.uk/2/hi/americas/country_profiles/1154116.stm date accessed 4 September 2012.

Bookman, M. A. and Bookman, K. R. (2007) *Medical Tourism in Developing Countries* (New York: Palgrave Macmillan).

Caribbean Export Development Agency (2011) *Final Report – Caribbean Health and Wellness Seminar* http://www.carib-export.com/website/SiteAssets/CEHEALTH REPORT REVISED.pdf date accessed 4 September 2012.

CARICOM (2011) 'History of the Caribbean Community (CARICOM)' *CARICOM* http://www.caricom.org/jsp/community/history.jsp?menu=community date accessed 4 September 2012.

Chambers, D. and McIntosh, B. (2008) 'Using authenticity to achieve competitive advantage in medical tourism in the English-speaking Caribbean' *Third World Quarterly* 29, 5, 919–37.

Clinical Research Management Inc. (2009) 'A minimal dataset for human resources in health in the Eastern Caribbean: Country report Barbados' *Regional Observatory of Human Resources in Health* http://www.observarh.org/fulltext/BAR_Database.pdf date accessed 4 September 2012.

Crooks, V. A., Kingsbury, P., Snyder, J. and Johnston, R. (2010) 'What is known about the patient's experience of medical tourism? A scoping review' *BMC Health Services Research* 10, 266.

Economic Commission for Latin America and Caribbean (2001) 'An analysis of economic and social development in Barbados: A model for small island developing states' http://www.eclac.org/publicaciones/xml/2/7812/G0652.html date accessed 4 September 2012.

European Commission (2007) *2007 Annual Operational Review: Barbados* http://ec.europa.eu/development/icenter/repository/jar06_bb_en.pdf date accessed 4 September 2012.

Foreign and Commonwealth Office (2010) *Barbados: Country Information* http://www.fco.gov.uk/en/travel-and-living-abroad/travel-advice-by-country/country-profile/north-central-america/barbados?profile=all date accessed 4 September 2012.

Gill, J. (2010, March 20) 'Health and wellness tourism a partnership with private sector, says Barbados Minister' *Barbados Government Information Service* http://www.gisbarbados.gov.bb/index.php?categoryid=13andp2_articleid=3439 date accessed 4 September 2012.

Gonzales, A., Brenzel, L. and Sancho, J. (2001) 'Health tourism and related services: Caribbean development and international trade' *Caribbean Export Development Agency* http://www.carib-export.com/SiteAssets/Health Tourism.pdf date accessed 4 September 2012.

Invest Barbados (2011, July 7) 'World-Class Hospital to Open in Barbados to Serve Global Medical Tourism Market' http://www.investbarbados.org/newsmain.php?view=Worldclass hospital to open in Barbados to serve global medical tourism market date accessed 4 September 2012.

Jessen, A. and Vignoles, C. (2004) *Barbados: Trade and Integration as a Strategy for Growth* (Washington, D.C.: Integration, Trade and Hemispheric Issues Division, Integration and Regional Programs Dept., Inter-American Development Bank).

Lashley, C. (2011, October 28) 'New hospital for Barbados' *Barbados Government Information Service* http://www.gisbarbados.gov.bb/index.php?categoryid=13andp2_articleid=6888 date accessed 4 September 2012.

Ministry of Tourism (2009, March) 'Annual Tourism Statistical Digest 2006' http://www.tourism.gov.bb/reports/2006_STATISTICAL_DIGEST.pdf date accessed 4 September 2012.

Paffhausen, A. L., Peguero, C. and Roche-Villarreal, L. (2010, March) 'Medical tourism: A survey' *Economic Commission for Latin America and Caribbean* http://www.eclac.cl/publicaciones/xml/7/39397/Medical_Tourism_A_Survey_L111_final.pdf date accessed 4 September 2012.

Pan American Health Organization (2001) 'Barbados' http://www.paho.org/english/sha/prflbar.htm date accessed 4 September 2012.

Pan American Health Organization (2005) 'Barbados: Health situation analysis and trends summary' http://www.paho.org/english/dd/ais/cp_052.htm date accessed 4 September 2012.

Pan American Health Organization (2007) 'Health in the Americas 2007: country profile for Barbados' http://www.paho.org/hia/archivosvol2/paisesing/Barbados%20English.pdf date accessed 4 September 2012.

Pan American Health Organization (2008) 'Health systems profile: Barbados' http://new.paho.org/hq/dmdocuments/2010/Health_System_Profile-Barbados_2008.pdf date accessed 4 September 2012.

Rodney, P. and Copeland, E. (2009) 'Safeguarding primary healthcare: A case study of Barbados' *Social Medicine* 4, 4, 204–8.

Walters, J., Fraser, H. S. and Alleyne, G. A. O. (1993) 'Use by visitors of the services of the Queen Elizabeth Hospital, Barbados, WI' *West Indian Medical Journal* 42, 13–17.

World Health Organization (2011) 'Global health observatory data repository' http://apps.who.int/ghodata date accessed 4 September 2012.

5

Tourists with Severe Disability

Angie Luther

In this chapter I will:

- Outline extant literature on barriers for physically disabled tourists while referencing conflicting models of disability and the experience of *impairment* as opposed to *disability*.
- Examine the physical condition of cervical spinal cord injury (C-SCI) and the most common medical complications associated with it.
- Illuminate the lived experiences and perceptions of individuals with C-SCI in relation to their physical impairment that might negatively impact participation.
- Highlight possible policy implications that could facilitate and/or enhance participation for individuals with C-SCI and other similarly disabled groups.

Introduction

Individuals with a severe disability still aspire to be tourists and to participate in holidays-taking, especially if they have acquired the condition in adulthood after they have accumulated tourist experiences. This chapter offers a detailed study of (non-)participation in holidays of severely physically disabled individuals with high-level cervical spinal cord injury (C-SCI). They require daily physical/medical assistance and so can be seen as 'patients' for the entirety of their current lives. Notwithstanding those seeking a 'cure' via stem cell therapy, who do not feature in this study, they differ, however, from those tourists who have become ill and are seeking a treatment – a medical intervention – to restore or improve their state of heath. Thus the chapter lays bare the multi-layered constraints that act upon tourists, or would-be tourists, who are also patients.

As some of the most socially excluded citizens, individuals with C-SCI are virtually absent from both tourism participation and enquiry. Extant literature – which is largely informed by the 'social model of disability' and predominantly centred on external barriers for tourists with general disabilities – has shed relatively little light on the realities of the situation of people with severe disabilities. According to Best (2007, p. 169), '[p]eople need to realise that not all is social construction, ideology or discourse, all people have bodies and our bodies are often fragile and feel pain'. With reference to empirical work that sought to enhance the position of individuals with C-SCI, this offering seeks, therefore, to explore the limitations of tourism to individuals with chronic health conditions by examining whether issues outside those widely cited external barriers equally inhibit, and/or exclude, them from participation.

Few members of the public, or even the general medical profession, have much knowledge or understanding of C-SCI, which involves paralysis in all four limbs (BASICS, 2005). Yet the condition known as *tetraplegia,* formally called *quadriplegia*, which is acquired either through disease (for example, muscular dystrophy, polio) or trauma (for example, road/sporting accidents, gunshot wounds), is one of the most devastating, both physically and psychologically (SIA, 2005). As such, the most commonly acknowledged emotions include anger, grief, fear, anxiety, loneliness, low self-esteem, hopelessness and despair at having lost control of every aspect of their lives (SIA, 2005).

Due to advances in specialist medical care, many people with C-SCI can now approach the lifespan of non-disabled individuals. Thus the central issue is no longer whether they can be kept alive, but has more to do with their quality of life. Apart from keeping 'healthy', psycho-social well-being is essential as there is a significant difference in life-satisfaction between individuals with and without physical disabilities, while those with severe disabilities are the least satisfied of all (Avis et al., 2005). Since holiday-taking is generally equated with an enhanced quality of life (Gilbert and Abdullah, 2004; McCabe, 2009), participation in holidays may offer disabled individuals greater overall life-satisfaction (Avis et al., 2005), particularly as they have the same travel motivations as the general population. These are predominantly socio-psychological in nature, such as a need for escape of routine and a desire for a feeling of freedom and general well-being (Shaw and Coles, 2004). Yet, despite the introduction of disability discrimination legislation, holiday-taking is still rare amongst disabled people and even rarer for severely disabled individuals (Darcy, 2002; Packer et al., 2007) who, although most in need, are the most unlikely to profit from its associated benefits (Gilbert and Abdullah; 2004; Card et al., 2006).

Holidays are often financially prohibitive for disabled people, at an estimated extra cost between 30 per cent and 200 per cent (Cameron et al., 2003) and even higher for those with severe disabilities (Burnett and Bender, 2001; Darcy, 2002). Furthermore, a multitude of other barriers continues to exclude many from participation (Shaw and Coles, 2004). Smith's (1987) seminal, theoretical study on barriers for disabled tourists categorised barriers into three groups: *environmental* (external to the tourist), *interactive* (the tourist's response to environmental barriers) and *intrinsic* (internal to the tourist and due to the tourist's own cognitive, psychological and physical function).

Increasingly informed by the politically correct social model of disability which argues that societal structures 'disable' individuals, not their physical limitations, the small amount of subsequent empirical work on tourism and disability issues has predominantly focused on environmental barriers encountered by *already*-participating tourists with minor to moderate disabilities. Moreover, it has largely centred on just two barriers within this category: staff attitudinal and physical accessibility barriers (for example, Card et al., 2006; Bi et al., 2007), advocating the removal of the latter essentially as a means to increase market share for the tourism industry rather than for social well-being (Shaw, 2007).

Exceptionally, a few studies have included one or two interactive barriers although, apparently, no internal ones. Daniels et al. (2005) reported that 'wheelchair manoeuvrability' was a skill essential to holiday success, although this reflected the *physical* capabilities of manual wheelchair-users which, again, relates very little to those with severe disabilities with power (electric) wheelchairs. Regarding 'communication challenges', Smith (1987) suggested that since communication is a two-way process, the non-disabled person is equally responsible for endeavouring to understand and respond appropriately to the person with the impairment. Research examining disabled tourists appears not to have addressed this barrier, although fear of foreign language problems when seeking medical assistance limited cancer patients' holiday plans to only English-speaking destinations (Hunter-Jones, 2004).

The few notable exceptions that include (already-participating) severely disabled tourists in their studies (for example, Burnett and Bender, 2001; Darcy, 2002; Packer et al., 2007) concluded that barriers for severely disabled people were both more numerous and complex. Hence, even though traumatic C-SCI is slightly more common than SCI (paraplegia), fewer people with C-SCI than with SCI are seen in public, tourism spaces included. In part, their non-participation stems from their 'buying into' the old but still highly influential medical model of disability which emphasises functioning limitations, and insists that it is the individual's physical *impairment* that prevents him/her from participating in mainstream society (Smith and Sparkes, 2008), tourism included. Despite such a politically incorrect notion, a more complete picture of the issues of (non-)participation might emerge if their experiences and perceptions of impairment were also included. After all, part of disabled people's lived experience is, according to Thomas (2004), linked to their level of corporeal impairment which is *not* socially engineered.

By questioning whether social factors are *wholly* responsible for disability (Morris, 1992; Thomas, 2004), the 'unmentionable' is mentioned in disability politics, that is, impairment or associated complications, pain or fatigue are, in some cases, also highly relevant to the experience of disability. As 'disability is, at some level, undeniably to do with pain or discomfort of bodies' (Williams, 1996, p. 206), Morris (1992, p. 10) laments the fact that there is a

> tendency within the social model of disability to deny the experience of our own bodies, insisting that our physical differences and restrictions are **entirely** socially created. While environmental barriers and social attitudes are a crucial part of our experience of disability – and do indeed disable us – to suggest that this is all there is to it is to deny the personal experience of physical or intellectual restrictions, of illness, of fear of dying.

On C-SCI, Humphrey (1994, p. 6) adds, 'the social model appears to have been constructed for healthy quadriplegics [as it] avoids mention of pain, medication or ill-health'. On a similar note, the 'Horrendous pain' felt by a former rugby-player with high-level C-SCI in Smith and Sparkes' (2008, p. 229) study is discredited by health-care professionals who do not recognise 'phantom limb pain' despite some neurophysiological basis for it. The authors go on to say that the belief these men have in their understanding of their own experience of disability, and of themselves, is consequently undermined and, according to Wendell (1996), they may even feel stigmatised, marginalised or socially oppressed by such a lack of recognition by others of their *physical* experience.

Smith and Sparkes (2008) also acknowledge the incapacity of the man with C-SCI to breathe without a ventilator, the pain and fatigue he feels and the fact that he has no control over his bladder or bowels. These are the very effects of *impairment* which, unlike disability, is not socially constructed (Thomas, 2004). What is more, by downplaying physical impairment to focus exclusively on political-structural barriers and a positive identity for disabled people, there is also the danger of representing disabled people as one homogenous group with the same needs (Smith and Sparkes, 2008). The following section on C-SCI, sourced from SIA (2005) and the medical education manual for staff where fieldwork for the study was conducted, will illustrate that this is clearly not the case.

The spinal cord is vitally important to the functioning of the body. Surrounded and protected within a column of 33 separate pieces of circular bones (*vertebrae*), it runs from the base of the brain, down the middle of the back, to around waist level, transmitting messages between the brain and different parts of the body in order to activate movement and sensation. Individuals may break their back or neck, but as long as the fragile, gelatinous spinal cord encased within the vertebrae is not damaged, they will be able to walk again once their bones have been stabilised.

In most cases, paralysis occurs when fractured vertebrae pinch the spinal cord, causing it to bruise, swell and partially, or completely, tear. The higher up the spinal cord the damage ('lesion') occurs towards the brain, the greater the degree of dysfunction. The most severe cases are those affecting the cervical (neck) region and are labelled C1–C7 depending on the position of the vertebra with which the lesion is associated, with C1 the very highest injury possible.

Many trauma patients above C4 die before being hospitalised as they are unable to breathe unassisted. Those who survive require a ventilator to breathe and assistance to properly cough to clear the chest and avoid infection. Unlike those with low-level SCI and thus upper body movement, assistance is also needed with all movement and personal tasks: washing, shaving, dressing, feeding and bladder and bowel management. A wide, non-folding and heavy-power wheelchair is used and controlled by a mouth or chin piece which can also manoeuvre and change the angle of the wheelchair seat to relieve skin pressure points avoiding skin sores, which, as will be revealed, is essential for maintaining health.

One of the world's most well-known individuals with C-SCI was the late actor Christopher Reeve. The former 'Superman' star was paralysed from the neck down in 1995 after he sustained a very high-level C-SCI (C2) from a horse-riding accident. As with so many formally highly active and adventurous young people, the painful irony was all too evident with Reeve. The man with the once super-human physical condition was unable to move below the neck, breathe without the aid of a ventilator or be left unattended 24 hours a day. Yet just 9 years after his injury, Reeve died at the age of 52, having succumbed to an infection from a skin sore which had spread to the rest of his body. This caused a heart attack, after which he fell into a coma and died the same day.

His untimely death illustrates the devastating consequences that such a common complication of C-SCI may have on anyone with the condition if someone like Reeve, with all his fame, fortune and round-the-clock care by resident medical staff in his home, could fall victim to the common skin sore. Thus, once an individual with C-SCI has been stabilised, careful attention to the prevention of potentially *life-threatening* medical complications plays a major role in the daily management of the condition.

The most common complication associated with C-SCI is the *skin/bed sore*. It can lead to amputation or death as in the case of Reeve. Other common and potentially fatal complications are *Autonomic Dysreflexia* (that is, an overreaction of the nervous system due to an irritation such as a full bladder, pressure sores or temperature changes), *respiratory problems*, *urinary problems*, *temperature regulation* and *circulatory problems*. With so much at stake, hospital rehabilitation programmes tend to focus on the medical condition rather than the person (Smith and Sparkes, 2008).

The effectiveness of such a rehabilitation process is, however, questionable as an early but one of the most comprehensive sociological studies of male wheelchair-users by Creek et al. (1989) inadvertently demonstrated. They found that the majority of the men were readmitted for periods to hospital due to skin sores or bladder problems. Further health problems occurred as non-specialist hospitals were unable to provide the necessary specialist medical and practical care, particularly for bladder and bowel programmes, to prevent secondary complications from occurring. The researchers proposed that the general medical profession's practical knowledge of SCI needed to improve, while Younis (1998) suggested that re-occurring medical complications lay more with the quality of rehabilitation programmes.

It emerged also that all forms of leisure for Creek et al.'s (1989) wheelchair-users were inhibited by factors such as higher levels of injury, fatigue, phantom limb pain, a dislike of being 'man-handled' and reliance on others to accompany them. Many also changed their leisure expectations because living life in a strict care-routine interfered with leisure activities, having, for example, to plan activities around bowel and bladder functions. The study further reported that the leisure activities of those with C-SCI were particularly restricted due to greater mobility and employment restrictions, reliance on carers and the type of leisure activity denied them as they were physically unable to play 'disabled sport', the most common leisure activity.

The following section presents the internal issues linked to non-participation that emerged from a qualitative study conducted in the US with participants with C-SCI. Data were collected from staff, clients and former clients of a specialist residential nursing home (The Home) for individuals with mostly high-level C-SCI. Methods included non-participant observation, informal discussions and in-depth, semi-structured and unstructured interviews with staff ('attendants') and (former) clients who had taken a holiday (self-defined as 'travellers') and who had not ('non-travellers'). While accessing any number of non-travellers presented no problems, the opposite was true for those who had taken holidays as travel for those with C-SCI was extremely rare. Thus, when data saturation occurred eliciting the same responses from non-travellers and from staff with no client-holiday experience, it was decided that the same numbers of non-travellers as travellers (six) would be used for individual interviews.

Interviewees were self-selecting, although a cross-section of client-volunteers was initially requested to ensure a range of the condition – especially individuals with very high-level C-SCI – was represented, including those with the condition through disease. This was achieved with participants' levels ranging from C1–C7 with the majority at C2 and ventilator-dependent. Consideration was also given, and achieved, to age, gender, ethnicity and socio-economic background, the latter including low-income individuals on state benefits and two wealthy travellers with multi-million dollar accident insurance settlements. Participants referred to clients and other people with C-SCI as 'quads' (from *quadriplegia*), and to ensure anonymity, all participants were given pseudonyms.

C-SCI and (non-)participation in holiday-taking

Without a distraction or a goal such as a holiday, the study found that most of The Home's clients became increasingly withdrawn and fearful about their condition. Some became particularly anxious and depressed about the 'hopelessness' of their situation which often culminated in daily pleas for assisted suicide. Yet, with the effects of the limitations of impairment instilled in them by the dominant medical model/discourse of disability during and following injury, virtually all clients believed holidays to be impossible for them, as 'Basically, the tourism market is for people who are working and walking, and that's not me' (Alex). They largely inferred that it was their impairment, rather than societal structures and attitudes, which prevented them from participation, as Jose implied when, looking down at his body and ventilator tube, he sighed: 'Like **this**, no....Tourism, I think it's not for me'. Yet, having insisted that holidays were not for 'the likes of him', his subsequent comment revealed, as with other non-travellers, an indirect yet subtle desire for holiday-taking when he explained: '[vacations are] just thoughts in my dreams... when I'm on a beach in Cancun [Mexico]'. However, despite the absence of a language barrier in Mexico for Spanish-speaking Jose, he added: 'I wouldn't feel comfortable so far away...and it'd be kind of scary' as there might not be a specialist C-SCI unit in the vicinity.

Traveller Ralf stressed how anxious clients were about health, and how they, and even most staff, held many misinformed views about clients' physical capabilities as 'most quads don't think they can fly, especially people on vents', which is not the case. Attempting to explain such ignorance with regard to their corporeal form, traveller Don commented:

> I've never seen an airline or a cruise commercial that has someone in a power [wheel] chair...so people expect they will be excluded just because they don't see anybody doing it. I've travelled enough that I know I can do it but most people in my situation...don't think it is very practical or possible.

Equally, staff and clients referred to the embarrassment, and thus the withdrawal from public spaces, that clients felt due to their physical limitations, such as being 'fed like a baby in public'; having strangers hear noisy suction machines bringing up their saliva and other secretions; and having their catheters changed in full view of others on flights. Yet amongst the most significant barriers to participation were a plethora of fears related to their 'fragile bodies' (Mia, director of The Home), ill-health and associated complications. Nevertheless, distinction must be made between having a severe medical condition like C-SCI – but still being 'healthy' – and being in poor health due to largely avoidable common side-effects of C-SCI, such as skin sores and urinary tract infections. At the time of interview, only the wealthy travellers were in poor health. Due to the potentially fatal consequences of a simple skin sore if not treated, one was confined to bed due to skin sores and had also been prevented from taking holidays in the past because of them. Likewise, the other had skin sores and, partly due to health problems, had not taken a holiday in six years.

Discussing fear, traveller Ralf explained, 'being in a situation where you're dependent on a mechanical device, fear is fairly constant'. Fear was a legitimate emotion for a multitude of reasons. Venturing out often meant being confronted by numerous access barriers since the most common misconception amongst service providers was that all wheelchair-users belong to one large homogenous group – that of the manual wheelchair-user. Yet, while a step might be a minor inconvenience for a manual wheelchair-user, but could be overcome, it was a complete barrier to access for the far wider, heavier power wheelchair which could not be tipped back. Consequently, when traveller Andrew arrived to find one step into his room at the pre-booked and only 'accessible' motel in the area, he and his attendants had nowhere else to go and so 'all ended up sleeping in the van that night'. Such accessibility issues are, according to Don, not just an inconvenience, but could also be very 'serious for your health if you get stranded somewhere overnight with no equipment', unable to recharge the wheelchair battery for its ventilator or to control the wheelchair to 'drive' and make weight-shifts to avoid potentially fatal skin sores.

For health reasons, travellers therefore insisted that 'you always have to find out on the internet and call ahead before you go' (Ralf) to confirm the accuracy of tourist information. Most tourists with general disabilities had simply checked websites or sometimes called ahead for the sake of convenience and comfort (Daniels et al., 2005). However, travellers' follow-up telephone calls were always necessary to verify any 'fully accessible' claims, as these largely applied only to manual wheelchairs. Mostly though, calls were made to request further details not provided in text or visible in photographs. However, when ventilator-dependent non-travellers tried to telephone for information, their attempts were frequently thwarted. Just as they were drawing breath to speak, the other person would put the telephone down believing the (speech) delay meant that there was no one on the line. After several failed attempts, non-travellers generally abandoned any further attempts to make essential telephone calls that would enable them to venture out without fear of accessibility barriers that might endanger health.

The physical act of travelling was frequently reported to have had a negative impact on participants' bodies. As 'most quads have a time limit for sitting up, and then they need to lie down' (nurse Eva), non-traveller

Len recalled the *physical* pain (Smith and Sparkes, 2008) felt during a three-day journey home in his grandfather's van: 'I was in agony [as] I sat in the wheelchair the whole time'. Most travellers with high-level C-SCI never contemplated long-haul flights because, as Ralf explained, sitting on a 'fourteen-hour flight would beat me up too much...My back, I can feel it...We are all different. We all have different kinds of feelings, sensations. Some quads don't have any; some have tactile sensations'. Since power wheelchairs are not permitted in airline cabins, even short-haul flights carried potential health risks because travellers were unable to manage their own weight-shifts. Consequently Ralf, and other travellers, admitted that when skin sores had developed during such flights, '[these vacations] leave a bad taste in my mouth [as] I get back in poor health'.

Being out of their wheelchairs on flights meant that even non-ventilator-dependent travellers needed to fly with assistants as 'I need help with weight-shifts;...I have no upper body strength, so the person next to me has to hold on to me to keep me from falling. Got a lap belt on that's only going to keep me off the floor, but [I] wouldn't want to fly without a care attendant' (Don). With no upper body movement and secured by a single lap belt – as opposed to a possible seven belts on a power wheelchair – their body could be thrust forward, or to the sides, when landing, taking off or during turbulence. Attendants therefore needed to hold onto the client's upper body because, if thrown forwards with head to knees, breathing could be affected. Travellers also sat on a special pressure-relief cushion placed on the aeroplane seat. This meant they were lifted so high up that even six-foot Jake's feet were left dangling, which led to skin becoming knocked and damaged.

Another potential threat to health was the fact that all travellers' wheelchairs usually got damaged or broken – '75 per cent of the time' in fact (Ralf) – in the luggage-hold. This added to their health concerns as Don explained: 'the airline loaned me a manual one which was no substitute...[as] I couldn't lay back and do [my] weight-shifts....I felt so glad to be home, never wanting to leave home again'. Despite the body's frailty and functioning limitations, health problems were nevertheless caused by airlines' inability to allow passengers to remain in the safety of their power wheelchairs to provide, on request, multiple security belts or equivalent replacements for damaged power wheelchairs.

Apart from the potential health risks caused by flights, extended journeys and inaccessible buildings, not only uneven, cobbled streets were difficult and uncomfortable to wheel over but also the 'bumpy' movement posed a potential health risk by possibly inducing spasms. The concern shown by traveller Andrew's comment, 'I've got my spasms under control with all the medication I take, but...I'm destroying my liver at the same time', clearly draws attention to the possibility of further underlying health issues associated with the taking of additional medicine necessitated, perhaps, by increased travel-induced spasms.

In addition to various daily medications, individuals with C-SCI require a large quantity of fresh water throughout the day for drinking and for their ventilators to keep the air moist for their lungs. Hence the potential health risk of purchasing unsafe drinking water was considered a major barrier to some international holidays. On one such trip, bottles of water believed to be mineral water were purchased, which had, in fact, been filled with contaminated tap water with seals glued back intact. Failing to distinguish between safe and contaminated water would have meant the client becoming ill and hospitalised and thus at risk of further medical complications occurring due to the widespread lack of practical knowledge of C-SCI still evident within general hospitals (Creek et al., 1989; BASICS, 2005).

Although all participants expressed health concerns, the fear of ill-health away from home was so overwhelming that it largely cancelled holiday-taking for non-travellers. Even when health was, or nearly was, affected on a rare day trip, non-travellers were put off further outings for a long time afterwards, and some even stopped going out altogether. For instance, Maria suffered the health consequences of over-heating when she was forced to stay out too long in the hot sun because her pre-booked 'Access' taxi to take her home from the coast arrived over three hours late. Despite sun screen on her skin, she got sun burnt and became very ill due to her fully paralysed body's inability to sense and deal with very hot or cold temperatures. This could well have developed into the potentially fatal complications of Autonomic Dysreflexia (that is, an overreaction of the nervous system due to an irritation such as temperature changes). As a result, she became extremely fearful about venturing anywhere outside again in case her health suffered as a consequence.

The fear of ill-health away from home also inhibited travellers and caused them great stress and anxiety during the extremely long holiday-planning process, always necessary to try to prevent 'all the things that could go wrong and affect your health' (Don). They knew only too well the health risks involved on the journey itself and whenever accessibility on arrival could not be guaranteed. Hence Jake's comment summed up the feelings of all travellers: 'I look forward to [a vacation], but it's tinged with fear and nerves of whether I might get sick'.

Equally anxious about clients' health were attendants. They tended to be reluctant to shoulder the extra responsibility of being away with clients with high-level C-SCI as they were especially medically vulnerable. This was because, if anything went seriously wrong, the client could become dangerously ill, or die, which also brought significant personal risks to their livelihoods as nursing licences could be suspended. Moreover, staff found having to 'playing mother' for the sake of clients' health particularly problematic on holiday as it could be met with great resistance, dampen high spirits and generally spoil the overall holiday atmosphere. Nevertheless, attendants insisted during a holiday that a client retire earlier to bed than desired to relieve skin pressure, or consume no more alcohol in case it affected medication. Such unpopular actions clearly conflict with the ideology of freedom and the right to do whatever one likes on holiday away from every-day constraints and routine – something that other disadvantaged groups had acknowledged as very important holiday benefits (McCabe, 2009). Even so, the effects of even a brief holiday brought long-term and light-changing benefits.

The most remarkable changes in attitude and lifestyle relate to the two travellers with C1, the most severe disability. Post-holiday, a new-found *raison d'être* emerged as, finally, they had 'proved to themselves they could have a good time as a quad and could do something that they hadn't even thought possible' (nurse Liv). A direct consequence of realising that it was actually possible to enjoy themselves as a severely disabled person was a desire to remain alive as neither traveller ever spoke again of suicide. For the first time since injury, the experience of holiday-taking had given them hope for the future simply because 'they [had] learned that it is still possible to enjoy life, even as a C1' (director of The Home). The motive then of individuals who continued to travel after acquiring C-SCI was to engage in a continuous rehabilitation process of 'normalisation' and re-integration in the pursuit of recovering 'health'/psycho-socio well-being. Travelling remained problematic, however, and strewn with health risks.

The following incident illustrates the even greater travel-induced health risks for ventilator-dependent travellers; it also highlights the necessity of a very costly three staff to one client ratio. When a fire broke out as a traveller was dining in a hotel's panoramic restaurant on the 16th floor, his power wheelchair, with attached ventilator, had to be abandoned in order for staff to carry him down flights of stairs. Two carried him sitting on their locked-together forearms leaning against them, while the third walked backwards manually pumping air into his lungs. Such a medically vulnerable person would not have survived without the assistance of these three trained attendants as he could never have survived being carried, as had been suggested, over a fire-fighter's shoulder – yet another indication of the widespread ignorance of C-SCI.

Requiring medical assistance in another part of the world where there was a language barrier would be even more concerning (Hunter-Jones, 2004) and hence, for Jose: 'a ten on a scale of one to ten for me *not* to go. See, when you're a walking person, you don't think about that. But in a wheelchair, you're worried about your health if you don't speak the language and are far away from home. That's our problem.' With similar concern, non-traveller Maria declared that if she ever ventured on holiday, it would be a short cruise, but only 'as long as they didn't just have a "general" doctor on board'.

Interestingly, while low-income travellers highlighted as a significant holiday barrier the very considerable extra costs of travel, spacious hotels and attendants for severely disabled tourists (Burnett and Bender, 2001; Darcy, 2002), non-travellers' over-riding financial concerns were always related to health in the form of insufficient funds to purchase travel medical insurance. With the state of clients' delicate health likely to deteriorate rapidly if, for example, skin became damaged during transit, most non-travellers were not willing to risk needing to go into hospital for treatment which they could not afford. Hence Susan confided, [although] 'I'd love, really love to go on vacation... it's impossible. I don't have the money for national medical insurance or even for insurance for [this] State.... Our biggest problem is medical insurance and we wouldn't dare go anywhere without it.'

In complete contrast, virtually no traveller had even considered travel insurance or health care beforehand, although both the wealthy travellers admitted that there 'definitely has to be a hospital in the vicinity. I usually call them up before I go about what kind of facilities they have and get a contact there at the hospital' (Ralf). Like the low-income travellers, neither would have any issue about going into a general hospital either. For them, however, this was because 'whenever I go to hospital..., I still take my [private] nurses... twenty-four hours' (Andrew) as 'my [private] nurses take care of me when I'm in... hospital' (Ralf). Low-income travellers took a completely different, however, as Lynette (with very low-level C-SCI) explained:

> It's all the same, any hospital... it doesn't matter. I can go to the hospital here which has a fabulous spinal cord rehab unit, but if you go to the Emergency Room, they say, they told me I'm not quadriplegic! I'm moving my arms, so I must be paraplegic.... They haven't even heard of certain terms to do with my disability! So you just have to speak up and say: 'This is what's happening to me, this is what I need'. You have to know your needs. You have to talk about the mattress you need... [and] say: 'You cannot leave me; I'm going to need to be turned. I'm going to need people to check on... my catheter, my urine out-put. I'm going to have to have a bowel programme'. If they are incompetent or refuse to do it, that's when it's really hard.

Likewise, Don reinforced the point that the responsibility for managing health always rested with the individual with C-SCI: 'even in the local specialist hospital here, I... have to let them know what to do – keep on at them. Specialist or general hospital, **you** have to be in control, manage your own situation.' He demonstrated this and his knowledge of a common complication on holiday in Mexico when he recognised that he had a urinary tract infection. As a result, he was able to instruct his non-medically trained Mexican attendant to acquire the correct antibiotics which cleared up the problem and the holiday continued.

Ironically, unaware of the incompetence and thus further potential health risks associated with most hospitals regarding C-SCI (Creek et al., 1989; BASICS, 2005), non-travellers could never contemplate travelling outside their neighbourhood or to any country whose language they did not speak. As with Hunter-Jones' (2004) cancer patients, this was because they feared communication problems at hospitals would be too great a health risk. Low-income travellers on the other hand ignored foreign language barriers since they understood that the art of persuasion and the power of their own knowledge of C-SCI, even if articulated via an interpreter, was their true safety net. In a similar vein, non-traveller Alex asserted, '[if] the person isn't familiar with their own care and directing their own care, health's going to depend completely on who's with them' and that person may not be as competent as hoped, as was the case with some private staff employed by the wealthy travellers who regularly suffered with skin sores. Thus, in contrast with travellers' desire simply to be tourists – albeit acknowledging their 'patient' status alongside – non-travellers' virtual abandonment of any hope of becoming tourists highlighted how they were reduced by their condition to remain solely patients.

Despite such varying responses to their condition, it is clear that the physical nature and function limitations of C-SCI, as well as (fear of) poor health and associated complications, were, for all participants, very real barriers or constraints to holiday participation. Albeit in a particularly extreme form, the travellers' accounts are also reminiscent of many of the fears, hopes, frustrations and obstacles experienced by other tourists with health issues. Furthermore, since the contemporary presumption is that travel is for 'the healthy', they break the mould of the 'healthy tourist'. Additionally, the study demonstrates the manifestation of the same factors that cause insurmountable barriers for those who do *not* travel. While the barriers initially appeared to wholly reflect Smith's (1987) internal barriers, on closer inspection they paint a far more complex picture of multi-layered and inter-connecting constraints. The most significant are highlighted below.

Conclusion

From the study it emerged that, despite an undisclosed desire, holiday-taking was virtually non-existent amongst individuals with C-SCI. This was largely because most believed that their level of corporeal impairment and/or associated issues excluded them. In short, physical limitations and (fear of) travel-induced poor health were largely perceived to be the root cause of restricted or total non-participation. Some of the issues that negatively affected participation were those potentially life-threatening medical complications that arose as a result of common side-effects associated with C-SCI.

Often these evolved from or were exacerbated by the choice of, or being stranded in, hot- or cold-weather holiday destinations; poor and over-long travel conditions; inadequate care in (general) hospitals; and the fact that most clients had not received the necessary instruction during hospital rehabilitation to equip them with the knowledge, skills, confidence and even inclination to take personal responsibility for directing their own health-care management. Thus, as Younis (1998) predicted, only when rehabilitation patients learn to fully understand their condition, and are properly *trained* to take charge of it by taking an active role in the daily management of it, can medical complications ever lessen to prevent an endless cycle of readmission to hospital. As such, rehabilitation programmes need to equip patients with better knowledge and skills to be able to take responsibility for their own health care in any given situation by giving precise and informed instructions. They need to explain clearly to others their personal and health-care requirements and routines, including manual evacuation of stools for which general hospital staff are often untrained.

Making fact-finding telephone calls for ventilator-dependent non-travellers also proved problematic largely due to tourism providers' lack of disability training and ability to recognise and take calls from individuals with delayed speech. Equally, this problem reflects a lack of instruction and practice during rehabilitation in making fact-finding, as well as emergency sip and puff, telephone calls. Thus, although delayed speech, like most of the other perceived internal barriers, appears to be uniquely 'internal', it was in fact eclipsed by environmental barriers that ultimately prevented participation.

Hence to facilitate holiday participation for individuals with C-SCI, hospital rehabilitation professionals need to design and provide practical workshops for patients to acquire the skills to communicate effectively with health-care professionals, attendants and tourism providers, and to take responsibility for directing their own daily health-care management. Moreover, given the lack of access to paid employment, policy makers need to provide adequate state benefits to fund the considerable extra cost of an annual holiday and medical/travel insurance necessary for those with high-level C-SCI. Finally, in acknowledging the existence of both structural truths and corporeal body truths that exist outside the mind (Thomas, 2004), it is hoped that this chapter has gone some way to demonstrate that:

- Physical barriers and attitudes in society, state policies and welfare support systems, as well as bodily differences and restrictions of impairment, *all*, impact on an individual's lived experience of disability.
- The corporeal experience of impairment differs according to the severity of the condition and the experience of impairment informs the way people with C-SCI live their lives and experience holiday (non-)participation.

- The perceived internal barriers of physical limitation, ill-health and life-threatening medical complications associated with C-SCI are complex, inter-related issues largely connected to and, at times, caused or exacerbated by environmental barriers; in a sense, albeit a more severe case, these also demonstrate the more generic experiences of those confronting the possibility of medical tourism.
- A major environmental barrier, and policy implication, relates to ineffective rehabilitation practices to impart the practical and communication skills necessary to self-manage health care, which predominantly affects an individual's ability to participate in tourism and, ultimately, to re-enter mainstream society.
- The experiences of those rare few individuals with C-SCI who travelled revealed unexpected and highly significant psycho-social consequences of how tourism participation impacted their desire to *live,* rather than continue to merely *exist,* as a disabled person, which consequently opened up opportunities previously deemed impossible.

References

Avis, A., Card, J. and Cole, S. (2005) 'Accessibility and attitudinal barriers encountered by travellers with physical disabilities' *Tourism Review International* 8, 239–48.

Best, S. (2007) 'The social construction of pain: An evaluation' *Disability and Society* 22, 161–71.

Bi, Y., Card, J. and Cole, S. (2007) 'Accessibility and attitudinal barriers encountered by Chinese travellers with physical disabilities' *International Journal of Tourism Research* 9, 205–16.

British Association of Spinal Cord Injury Specialists (2005) 'Patient survey' http://www.basics.pwp.blueyonder.co.uk/patient_suvey.htm date accessed 4 June 2007.

Burnett, J. and Bender, H. (2001) 'Assessing the travel-related behaviors of the mobility-disabled consumer' *Journal of Travel Research* 40, 4–11.

Cameron, B., Darcy, S. and Foggin, E. (2003) 'Barrier-free tourism for people with disabilities in the Asian and Pacific region' http://unescap.org/ttdw/Publications/TPTS_pubs/pub_2316/pub_2316_tor.pdf date accessed 15 November 2005.

Card, J., Cole, S. and Humphrey, A. (2006) 'A comparison of the accessibility and attitudinal barriers model: Travel providers and travelers with physical disabilities' *Asia Pacific Journal of Tourism Research* 11, 2, 161–75.

Creek, G., Moore, M., Oliver, M., Salisbury, V., Silver, J. and Zarb, G. (1989) *Personal and Social Implications of Spinal Cord Injury: A Retrospective Study* (London: Thames Polytechnic).

Daniels, M., Rodgers, E. and Wiggins, B. (2005) 'Travel Tales: an interpretive analysis of constraints and negotiations to pleasure travel as experienced by persons with physical disabilities' *Tourism Management* 26, 919–30.

Darcy, S. (2002) 'Marginalised participation: physical disability, high support needs and tourism' *Journal of Hospitality and Tourism Management* 9, 4, 61–72.

Gilbert, D. and Abdullah, J. (2004) 'Holidaying and the sense of well-being' *Annals of Tourism Research* 31, 103–21.

Humphrey, R. (1994) *Thoughts on Disability Arts*, DAM 4, 1.

Hunter-Jones, P. (2004) 'Young people, holiday taking and cancer: An exploratory analysis' *Tourism Management* 25, 249–58.
McCabe, S. (2009) 'Who needs a holiday? Evaluating social tourism' *Annals of Tourism Research* 36, 66–88.
Morris, J. (1992) 'Personal and political: A feminist perspective on researching physical disability' *Disability, Handicap and Society* 7, 2, 157–66.
Packer, T. L., McKercher, B. and Yau, M. K. (2007) 'Understanding the complex interplay between tourism, disability and environmental contexts' *Disability and Rehabilitation* 29, 4, 281–92.
Shaw, G. (2007) 'Disability legislation and the empowerment of tourists with the empowerment of tourists with disabilities' in A. Church and T. Coles (eds) *Tourism, Power and Space* (London: Routledge).
Shaw, G. and Coles, T. E. (2004) 'Disability, holiday making and the tourism industry in the UK: a preliminary survey' *Tourism Management* 25, 3, 397–403.
Smith, R. W. (1987) 'Leisure of disabled tourists: Barriers to participation' *Annals of Tourism Research* 14, 3, 376–89.
Smith, B. and Sparkes, R. (2008) 'Narrative and its potential contribution to disability studies' *Disability and Society* 23, 1, 17–28.
Spinal Injuries Association (2005) http://www.spinal.co.uk/ date accessed 8 December 2005.
Thomas, C. (2004) 'Disability and impairment' in J. Swain, C. Barnes, S. French and C. Thomas (eds) *Disabling Barriers – Enabling Environments,* 2nd edn (London: Sage).
Wendell, S. (1996) *The Rejected Body* (London: Routledge).
Williams, G. H. (1996) 'Representing disability: some questions of phenomenology and politics' in C. Barnes and G. Mercer (eds) *Exploring the Divide: Illness and Disability* (Leeds: Disability Press).
Younis, A. (1998) *The Social Reintegration of Patients with Spinal Cord Injury in Palestine.* PhD thesis (unpublished) (Ulster: Ulster University).

Part II

Patients as Tourists

6

Beauty and the Beach: Mapping Cosmetic Surgery Tourism

Ruth Holliday, Kate Hardy, David Bell, Emily Hunter, Meredith Jones, Elspeth Probyn and Jacqueline Sanchez Taylor

In this chapter we will:

- contextualise cosmetic surgery tourism and sketch some of its defining features;
- look more closely at how cosmetic surgery tourism works as a phenomenon that assembles a complex set of people, places and practices;
- examine how the cosmetic surgery tourism industry is developing;
- consider debates in tourism studies to understand what it means to call our subject cosmetic surgery *tourism*.

Introduction

Cosmetic surgery tourism – traditionally defined as the movement of patients from one location to another to undertake 'aesthetic' medical procedures – is a significant and growing area of medical tourism (Reisman, 2010). The UK's annual International Passenger Survey, produced by the Office for National Statistics, shows that approximately 100,000 UK citizens go abroad each year for medical treatment (a number rising by about 20 per cent annually), and cosmetic surgery tourism is estimated to make up about 85 per cent of the medical tourism market in Australia (Connell, 2006). It has also been suggested that although financial crises, privatisation and the rising cost of health care may have slowed the demand for cosmetic surgery in some 'developed' countries, crossing national borders to procure those surgeries appears to be increasing as consumers seek out low-cost procedures abroad (see Bell et al., 2011). The industry itself is acquiring institutional 'thickness' as the various agencies and agents involved increasingly coalesce into assemblages, regulatory and promotional bodies, financial regimes, and complex flows of bodies, knowledge, technologies, money, ideas and images. As Mainil et al. (2010, p. 749) summarise, 'the global network society has touched the medical field and there is no going back'.

While some commentators argue that 'tourism' is an inappropriate label to apply to practices of travelling to obtain medical treatment (Glinos et al., 2010; Kangas, 2010), we want to set that debate aside and instead look more closely at *how cosmetic surgery tourism works*: this chapter draws on a large-scale, multi-site, mixed methods research project exploring the practices, sites and experiences of cosmetic surgery tourism. In particular, in this chapter we are interested in offering an analysis of the issue of place within cosmetic surgery tourism, in terms of both image and experience. But before we focus on this discussion, we want to briefly contextualise cosmetic surgery tourism and sketch some of its defining features.

Medical mobilities

Cosmetic surgery tourism can be understood as part of a larger trend towards health, wellness or medical tourism, in which tourists partake of various health-benefiting practices, from 'detox' to hip replacements, and thus it is one important element in the globalisation or disembedding of health care (as this volume shows; see also Reisman, 2010; Connell, 2011; Cockerham and Cockerham, 2012). Medical tourism has come to be generally defined as travel-by-choice from a 'developed' country to access health-care services in a 'less wealthy' country (Jones, 2011), although the realities of the actual flows between 'home' and destination countries are much more variegated and non-isomorphic (Glinos et al., 2010). There is an emerging body of academic research focused on this set of practices, including work taking a tourism management/industry perspective (Goodrich and Goodrich, 1987; Connell, 2006, 2011), some by health-care researchers (Castonguay and Brown, 1993; Ramirez de Arellano, 2007), and a little by sociological and cultural researchers, often as part of broader considerations of cosmetic surgery (Elliott, 2008; Jones, 2008, 2011). There has also been work on the implications for the health-care systems 'back home' and on medical insurance, including commentary from clinicians concerned with the burden of aftercare once tourists return (Birch et al., 2007; Burkett 2007). We will draw on some of this work in our discussion. In addition to the academic literature, there is abundant press coverage and a 'how to' literature, plus many websites, discussion forums and blogs (Mainil et al., 2011). The Internet has, in fact, played a prominent role in how the practices, places and industry have developed – it might be argued that without the Internet, the industry as such would not exist (Lunt et al., 2010).

Accounts of the development and current form of the cosmetic surgery market often highlight four drivers of its growth: cost, access, quality and service (Schult, 2006). In countries with private health care, cost can be a very powerful motive – there is considerable global variation in the cost of procedures, a variation that tourists are (sometimes) able to exploit. However, in countries with public health-care systems, waiting lists and quality of care can be more important than bare cost. Also, in some places with nationalised care, some procedures are designated as 'elective' and so not provided freely, making cost a factor in these cases, too. Access does not simply mean proximity in a straightforward geographical sense, though some studies emphasise tourism between neighbouring countries or across adjoining borders (Glinos et al., 2010; Kangas, 2010). More commonly, access is more an issue of mobility and the convenience of travel – the low-cost airlines have played a major role in reshaping this geography of proximity, and airlines and airports are key nodes in these emerging networks. Access might also involve familiarity or 'cultural proximity' – language, for example (guidebooks often reassure readers that staff speak excellent English) – and familiar destinations (Glinos et al., 2010). 'Cultural proximity' can also extend to a vaguer notion of the kinds of destinations that are *imaginable* as offering appropriate levels of service – particular places trade on certain images (or myths) in order to 'sell' themselves as a cosmetic surgery destination (Holliday et al., 2013). Quality and service are factors played up in promotional materials, as might be expected. Plush hospitals/hotels, standards of care, training of surgeons (Western training is often emphasised) and aftercare are all key elements of service. Facilities targeting British travellers sometimes compare their high-quality facilities to the UK's 'third world' National Health Service, promising much higher standards of care than are available at home.

It is also important to consider the extent to which cosmetic surgery tourism is promoted *as tourism*, notwithstanding debates about the appropriateness of this moniker. Reading the spaces, practices and experiences through a tourism lens, we would argue, opens up productive conceptual and analytical avenues (Mainil et al., 2010; Bell et al., 2011). For example, it is uncontroversial to suggest that tourism happens in the imagination as well as in actual lived experience (Urry, 1990, 1995; Rojek, 1997), and also to say that the holiday is often constructed as a time and place apart from 'ordinary life' (though see Edensor, 2001). Where we choose to go on holiday, including where we choose to go for surgery, draws upon those ideas and images we already have about different places and cultures, and conjures ideas about the experiences we expect to have there. So the 'place myths' of destinations may be central to choices about where to have surgery – a prominent example being Brazil, known for its obsession with beauty and therefore for its exacting standards of aesthetic perfection (Edmonds, 2007, 2010). Guidebooks and websites draw on these 'place myths' and in doing so produce an interesting and geographically variegated set of cosmetic surgery tourism offerings, and tourists select their destinations *at least in part* as particular places where they not only expect certain prices and levels of service but also imagine an encounter with a particular host culture – including, but certainly not limited to, its surgical culture. Packages offer surgery plus recuperation in a beautiful resort location or specialised retreat, while also offering more familiar tourist experiences for patients and accompanying family members or fellow-travellers, such as sightseeing or even safaris. There's something very interesting – and very successful – about this packaging, and about the way it utilises ideas about holidaying as a particular kind of experience, even when the holiday is centred on a facelift.

Cosmetic surgery *tourism?*

Embedded in the preceding discussion has been the issue of the holiday as a time (and space) apart – as a distinct demarcating of 'me time' (or 'us time' – since commentators note the high number of couples who travel together and have procedures together). How can we conjoin this idea of the restful, restorative and self-indulgent holiday with the actual practices of undergoing surgery, with all the pain and trauma, stress, and discomfort? Does the actual act of combining something that is at once dreaded *and* desired, with the practices of holidaying, mark the surgery as more eventful or more bearable, more exceptional or more mundane? John Urry (1990) once defined tourism as entailing temporary journeys to places offering out-of-the-ordinary experiences. While some commentators argue that holidays should not be simply seen as a break from 'normal' daily life, but more a continuation of it (Edensor, 2001), we would argue that the idea of the holiday as a time and space apart remains a powerful feature of the ways in which tourism experiences are imagined (and, indeed, experienced). The holiday carries a weight of expectation: that it will be out-of-the-ordinary, and in some way either restorative or transformative (or both). The tourism industry reproduces this idea, which also circulates in popular culture and everyday talk (Mainil et al., 2011). At this level, there is no dissonance between the terms 'cosmetic surgery' and 'tourism', since elective cosmetic surgery is itself framed as a 'treat' or 'reward', something well earned and that will also repay its investment by bringing into being a 'new you' (Gimlin, 2007), and also as a transformative experience – albeit of the body not the soul. Cosmetic surgery and 'foreign' holidays are in essence both part of the same economy of making and marking successful selves, self-improving selves: travel and surgery are both forms of cultural capital.

Locating cosmetic surgery tourism

That said, a frequent focus in the academic literature on cosmetic surgery tourism homes in on *placelessness*: cosmetic surgery tourism is depicted as a practice for transnational elites whose lifestyles ride the 'global flows' but who, as a consequence, are variously characterised as footloose or rootless – the 'nowhere men' (and women) in Pico Iyer's (1997) description. Also tracking this class, with an eye on their cosmetic practices, is Anthony Elliott, who explores the fictional experiences of 'typical' tourists Sharyn (owner of a hairdressing chain) and Grant (a lawyer) from London, who travel to Malaysia for a breast augmentation and liposuction respectively. Elliott is keen to point out the enclavic nature of luxurious hospitals located in otherwise modest economies and the power relations of mobility versus immobility that cosmetic surgery tourism produces. Drawing on Augé (1995), Elliott represents these hospitals as 'non-places' of 'recuperation' that are 'placeless, indistinguishable and indistinct' (Elliott, 2008, p. 104) and the tourist experience as a 'solitary ordeal' (p. 105) which nevertheless provides the patient with a 'springboard to reconfigure mind and body' (p. 107) before returning to everyday life. But this logic of place-specific travel only makes sense to precisely those footloose transnational elites; other tourists are travelling to different places, driven by different motives. If the foregoing discussion centres on those who occupy business class seats, that which follows is more focused on the economy class traveller. As Glinos et al., (2010, p. 1153) write, 'while wealthy patients seeking the best care have always travelled, a new kind of patient is now travelling to save money. This implies a change not only in the purpose of patient mobility but also in the direction of flows.' It also, as our empirical research has uncovered, reveals very different understandings of place and placelessness.

It is, however, in order to attract footloose globetrotters that cosmetic surgery tourism promotional materials often foreground a very 'placeful' experience – conjuring images of the specificities of place as a vital component of the package on offer. The cosmetic surgery tourism industry has therefore invested heavily in producing place images that provide attractive motifs. Recurrent images include high-tech and brand-new facilities, smiling nurses and beautiful bodies; websites and promotional materials produce well-crafted juxtapositions of the surgical and the touristic (Holliday et al., 2013). As Erynn Casanova (2007, p. 11) writes, these materials most frequently feature 'the ubiquitous image is of a woman in a bikini lounging on the beach'. Aside from generic motifs, particular places also trade on their *distinctive* images, highlighting their uniqueness as a tourist destination.

Ackerman's (2010) ethnographic research in Costa Rica offers an exemplary example of the marketing of a whole host of place images/myths in order to sell a particular tourist destination. Ackerman is particularly interested in the development of what she calls 'recovery retreats'; somewhat like Elliott's 'restorative nonplaces', these are 'a hybrid of hotel, home, and clinic, and generally offer Internet access, America-style food, US cable TV, social activities, transportation to and from clinics, English-speaking nurses, exam rooms for physicians, lush gardens' (Ackerman, 2010, p. 408). In these carefully manicured places, tourists from the US recuperate in an environment that blends state-of-the-art medical care with notions of Costa Rica as a tourist paradise and essentialised ideas of the 'caring nature' of local healthcare workers. Drawing on earlier campaigns promoting the country as an ecotourism destination, cosmetic surgery tourism place images emphasise the restorative capacity of the place itself – a healing landscape – mixing this with, as Ackerman writes, a 'potent blend of colonial, utopian and technological imaginaries' (p. 412). Promotional images balance certain motifs of exoticness with those of naturalness and those of high-quality, high-tech modern facilities. In particular, the quality of care provided is projected as an inherent characteristic of the Costa Rican people, and therefore as place-specific. Place and surgery are here packaged together in ways that seek to highlight and promote what is distinctive about the destination and the cosmetic surgery tourism experience – a kind of surgical *terroir*.

From our research it is clear that clinics use very specific forms of address and constructions of place in an attempt to market and locate themselves in relation to positive place myths and well-trodden paths of prior tourist migrations. The following section explores some themes that emerged from a website analysis we conducted on three of the locations from the broader study – Spain, Thailand and the Czech Republic.

Tourist snapshots

With its glorious climate, spectacular scenery and fine beaches, Spain has long been one of Britain's favourite destinations – for holidays, second homes and retirement. Now, the emergence of high-quality clinics – coupled with favourable prices – is making Spain important as a medical destination as well. The large expat population in Spain means that every kind of English-speaking facility is readily available, and the number of low-cost flights from all British airports makes getting there easier than to almost any other European country (http://netdoctor.privatehealth.co.uk/private-dentistry/cosmetic-dentistry-abroad/spain/).

This description of Spain as comprising luxurious landscapes, familiarity, high quality and easy accessibility – both geographically and due to the predominance of English – offers a useful summary of the imagery and identity represented within cosmetic surgery tourism websites. Spanish websites evoke images of luxury and relaxation, with home pages (in English) displaying beach resorts with up-market connotations such as expensive-looking yachts and pleasure cruisers. These images serve to connect cosmetic surgery with beauty and success, specifically in the form of wealth.

'Perfect' (although distinctly un-augmented) bodies populate homepages. Text connects 'youth', 'health' and 'contemporary image' with beauty and addresses patients who want to look 'beautiful and attractive while enjoying the psychological benefits of improved self-esteem and renewed confidence' (for example, http://www.oceanclinic.net/plastic-surgery/). This will be delivered, the text promises, in superb and tranquil premises in beautiful Marbella. The beauty of the landscape is thereby connected with the beautiful bodies produced by the clinic. Representations of surgeons as 'whole' often contrast with the dismembered women's bodies representing patients.

And both the smooth bodies and smooth surgeons jar with the before-and-after shots of cosmetic surgery. After the serenity and perfection of the generic images, these photos of actual patients are not only shocking but grossly interruptive of the surgical fantasy. We are brought quickly down to earth by 70-year-old patients face-lifted to look 50 and middle-aged men with their ears pinned back.

Luxury is also clearly operationalised on Thai websites, albeit in a distinctive way. Somnio Medical[1] poses the question: 'What better way to recover than in a 4–5 star accommodation? When the only thing you will need to do is pick up the phone and have your pillows and meals delivered (and have someone else do the housework daily!)'.[2] Within the Thai websites we observed considerably more range, with some sites far more 'amateur'[3] and others distinctly more sophisticated. However, almost all include images of beaches, usually tropicalised through the use of images of long boats and palm trees and tanned, slim, youthful women in bikinis, and some men – especially male torsos. Also present, although less common, are images of groups of women poolside or shopping, women's and men's body parts in underwear, happy couples (always heterosexual) and surgeons (always men in white coats in a clinical setting). There are scattered images of actual surgeries, of Thailand's urban areas and of Thai people who are not medical professionals.

The lotus flower is nearly ubiquitous. Many sites also include images of massage and yoga, and pictures that evoke an 'Eastern spiritual' mood: faces with eyes softly closed in meditation, feet resting in pools of flowers, hands in the *wai* (prayer) position, Buddhist statues and temples. This holistic, ethereal, 'zen' representation is reflected in the first paragraph that a potential tourist encounters on the Somnio website:

> In Latin Somnio means 'to dream' and at Somnio International Medical Holidays®, we take the guess-work and hard work out of assisting you in achieving your dream.

This rather ethereal framing sits uneasily, perhaps, with a large table comparing prices for different procedures in Thailand and Australia. The front page also emphasises more material advantages, such as 'world-class hospitals, internationally trained surgeons and doctors [and] state of the art medical facilities'.

Although value for money is emphasised, Thailand is described as an inviting, sensual, peaceful environment, promising 'dedication and "genuine" care from the doctors', including 'a friendly "Thai" smile on every corner'. Key to cosmetic surgery marketing in Thailand ('the land of smiles') is the notion of care, notably the idea of 'genuine', 'sincere' or 'authentic' care from medical professionals. This, combined with the notion of a carefree, sensual holiday with a notable 'spiritual' element, forms the basis of advertising cosmetic surgery tourism in Thailand.

In both Spain and Thailand, then, cosmetic surgery is connected to the idea of a journey or transformation. While this is not unusual in either regular tourism brochures or the academic tourist literature (see debates in Edensor, 2007), the transformative experiences of tourism are most often framed as mental or spiritual ones. While Spanish clinics frame cosmetic surgery as a 'journey' to a new and better 'you', this is located very much in the material – transforming the body produces a transformed self, and this new self is connected with success in the form of material wealth. Luxury and individual class mobility are presented as the *outcome* of cosmetic surgery tourism to Spain. In contrast, despite the new age/spiritual feel of Thai websites, the foregrounding of price on the websites represents luxury as something that can be purchased as part of the *process* of having cosmetic surgery in Thailand. Both Spain and Thailand foreground the beach and the sun as spaces of healing (despite the dangers of exposing fresh scars to the sun's UV rays) – beaches evoke the idea of holidays and relaxation and offset worries about risks and pain of surgeries. They provide an immediately recognisable and standardised restorative location. Nevertheless, the distinctive features of specific places are also mobilised and articulated as healing properties of particular landscapes.

Places of skill

By contrast, cosmetic surgery tourism websites in the Czech Republic depart markedly from those of their Spanish counterparts. Most noticeably, in place of images, the home pages are dominated by text, prices are foregrounded and body parts are conspicuous by their absence. Websites feature images such as clinic exteriors, patient consultations, surgeries in process (although the patient is hidden under scrubs and operating sheets) and nightscapes of Prague. The text promises 'highly qualified plastic and aesthetic surgeons with many years of experience in the field of plastic and aesthetic surgery in Canada and the USA'. Surgical skill is prioritised and evidenced by reference to other surgical destinations – Canada, the US, Austria and the UK. Hygiene is also given prominence: one website promises 'supra-septic conditions, disposable operating props, hygienic surgical dressing, absorbent suture' – reassuring perhaps, but also startlingly graphic.

There is also a clear attempt to establish Prague as a key site in cosmetic surgery culture. For example, Beautiful Beings website positions Prague as the 'cosmetic capital' of Europe. The Czech Republic has a long tradition of plastic surgery; the Czech professor Francis Burian is regarded as one of the pioneers in the field of plastic surgery. Charles University, Prague, was home to Europe's first faculty for plastic surgery which opened in 1920.

Alex Edmonds (2010) shows how cosmetic surgery can be an important symbol of modernity in newly developed countries. In Brazil, the cosmetic surgeon Ivo Pitanguy has become a figure of national pride, carnival songs eulogising his role in bringing beauty to ordinary (poor) Brazilians. Cosmetic surgery is a technology, and being at its forefront brings value. High levels of surgical skill are stressed both through the 'relational geographies' of referring to other places – 'first world' countries where, we are supposed to assume, standards are highest – and also through the construction of histories of indigenous expertise and cosmetic surgery foundation myths. Much like the Spanish and Czech websites, the 'international' experience of the surgeons is stressed in Thai websites, along with the importance of the regulation of the hospital (it conforms to the American JCI system – 'one of the most advanced and demanding accreditation systems in the world').

At the same time, however, other places are found deficient. Czech and Thai websites stress that clinics are 'state of the art' and that cosmetic surgery and anti-ageing medicine are practiced in the same theatres as medically necessary heart transplants, for instance. Very low infection rates are contrasted with 'superbugs' in hospitals 'back home' in Australia and the UK. Stringent industry regulation is constructed as better than the lax regulation of most other countries in the world.

The place of the clinic

While the websites discussed above foreground the locatedness of clinics and hospitals and draw on the typical tropes of tourist literature to create a specific meaning of place related to images of rest, healing and recuperation, the reality of clinics in our research has sometimes been rather different. In Thailand, clinics are fairly embedded in particular locations – Bangkok, or sites such as Phuket, that has become, in the last 10–20 years, part of the mainstream Australian tourist imaginary as a low-cost beachside resort sold in part via a fairly essentialised notion of Thai hospitality (which the cosmetic surgery tourism industry also trades on). And travellers from Australia to Thailand tread a very well-worn cosmetic surgery tourist trail, in search of cut-price but good-quality cosmetic surgery – though even at the comparatively low rates of accommodation and food in Thailand, the cost of surgery is still significant for the tourists we encountered, who therefore managed their budgets carefully. Many Australians use upmarket hospitals in both Phuket and Bangkok, though a significant number use group tours to budget hospitals lacking the fancy trimmings (lounges, artworks, fast food outlets) that make the upmarket hospitals more reminiscent of luxury hotels. For Australian travellers, the shortest journey time on this trail is Perth to Phuket, which is only four and a half hours, and this is one reason why there is such a boom in the Western Australian market. From the Eastern states, however, the flight is between seven and nine hours, although several of our participants chose budget airlines with two stopovers that took them closer to 20 hours. The minimum recommended time for a stay was between 7 and 10 days, rising to 15 days for multiple procedures, and most people we met followed this surgical advice, some interspersing their recovery with shopping trips, cookery courses or visits to 'ping pong' bars on the margins of the sex tourism industry.

Most of our participants had also left young families at home, usually with their partner (if they travelled alone or with a girlfriend), but sometimes with a parent (usually if they travelled with their partner). They wanted to return to their families as soon as possible because they missed them so much, and because their partners usually had a limited amount of leave from work. Some also talked about the burden of leaving children with family members and wanted to return quickly so as to remove this burden. (Some male partners had never been left at home with the kids before, so this burden extended to them.) This point is also related to cost: many people we met thought about bringing their whole family with them for a holiday, but could not afford to do this. The alternative was to go alone or with one support person and find alternative care for the children during the time away, which was therefore planned to be as short as possible. In addition, to offset cosmetic surgery as 'selfish individualism', patients tried to spend as little as possible and to get back to relieve helpers as quickly as they could.

Our research in Europe also mapped an enterprise significantly different to the promotional image found on websites or the elite hedonism described in some accounts. Because of the interlinked business interests and currencies of the European economies, the Eurozone crisis appears to have had a significant impact on European cosmetic surgery. Once an extremely popular destination, currency fluctuations between the pound and the Euro have increased the relative costs of surgeries in Spain, reducing it from an extremely popular cosmetic surgery destination to a much more limited market. While Spain remains popular for British patients, they are drawn mainly from the one million strong expatriate population that already lives there. Spain also appears to be popular with wealthy Russians, many of whom have holiday homes in Marbella, where the majority of the clinics that advertise in English are based. While Spanish websites acknowledge that cost savings at Spanish clinics are modest compared to the UK, other benefits of the particular location are strongly highlighted – beaches, sun, good food. Ironically, however, this strong construction of space belies a rather more virtual culture of temporary transnational medical assemblages (Mainil et al., 2010).

For example, our research in Marbella focused primarily on two clinics. The first, an up-market clinic, comprised a breakaway group from a larger cosmetic surgery tourism organisation who felt exploited by their previous employer. The clinic rented an operating theatre and recovery rooms in a private hospital with ten beds set in perfectly manicured gardens in the former mansion of a famous Hollywood film star. The hospital was shared with other clinics specialising in mental health and oncology. The surgeon, who was Italian, travelled to Marbella from Madrid for several days each month, performing surgeries, then follow-up appointments with patients. Patients were roughly one-third Spanish, one-third British ex-pats and one-third Gibraltans – shopworkers, hairdressers, retired ex-pats – who wanted to regain 'lost bodies' 'spoiled' by pregnancy, caesarean, dramatic weight loss or ageing. The surgeon was joined by an Irish intensive-care nurse, a British patient co-ordinator and a Spanish receptionist. Such a clinic is theoretically highly mobile and can re-group almost anywhere in Europe, depending on where currency rates provide the best bargains and the biggest business. However, the location of the clinic in Marbella is at once very important because it has built strong local reputation – almost all of the patients we have interviewed so far select surgeons through word of mouth, albeit sometimes electronically. So a clinic's most effective marketing strategy is personal recommendation – but by the same token, when a surgery goes wrong, the clinic's Marbellan 'surgical community' is quickly informed. So location is both an opportunity and constraint.

In Tunisia, the globalised mobilities of medical staff were even starker. Marketed through a British agent, the website again deploys the relational geographies of skill – 'Doctors trained to high French standards, GMC Specialist Registered surgeon (UK) with consultations in London, Board Certified surgeons from the USA'. Patients travelled from the UK and France. The French surgeon practiced mostly in Paris, but was offering cut-price deals for French patients in Tunis, where hospital costs were far lower. The agent identified this surgeon as potentially offering good value for UK patients too and began marketing it as a destination. We visited the hospital in February 2012 along with the agent and three British patients. Two cosmetic surgeons were working in the hospital there, one from Paris another from the US. One left the country at the end of his visit with suitcases full of Euros – it was normal in many clinics for patients to pay in cash, though some took credit cards. The hospital had three major patient groups – European medical tourists, wealthy Tunisian private patients and war veterans and ordinary people who had been caught in the cross-fire of the Libyan civil war. Who could have known that our research would be caught up not just in the Eurozone currency crisis but also the Arab Spring?

The patients we travelled with were largely unaware of where Tunisia was and were surprised to find it was in Africa, which they had previously thought was a country, not a continent. Patients returned home as soon as they were well enough to travel – there was no 'holiday' attached to their medical tourism. One patient had so much cosmetic surgery she was extremely disoriented and left to recover in a side ward with care staff who spoke little or no English. Another was surprised at the mix of patients in the hospital and commented that they 'looked like they'd been in the wars', especially when one Libyan, clearly highly distressed, attempted suicide in the hospital lobby. The final patient, who had a blepharoplasty, befriended other patients of differing nationalities and medical conditions, and reflected that she felt rather ashamed coming for a 'beauty treatment' when others in the hospital needed surgery to save their lives. From our fieldwork, therefore, a different set of experiences, journeys and geographies becomes apparent, countering the notion of a particular articulation of placelessness in cosmetic surgery tourism.

Discussion

What, then, are we to make of this in relation to the literature on cosmetic surgery tourism? Beginning this project we had expected to see high levels of patient mobility, with savvy people travelling to secure the best bargains. We were prepared to witness the exploitation of medical staff in low-wage economies by global elites – like Elliott's (2008) cosmopolitans sampling enclavic and recuperative non-places, but remaining untouched by the 'real' place. In contrast to Elliott, we found an industry dominated by 'ordinary people' – working-class or lower middle-class hairdressers, beauticians, lap-dancers, shopworkers, taxi-drivers and so on. In some places it was possible, because of differential currency rates, to 'live rich' even for these patients. One woman in Tunisia was delighted to acquire a designer handbag at an unbelievably low price – a price equivalent to a month's salary for a nurse working in the hospital that treated her. But currency differentials don't automatically equate to exploitation. What was surprising, perhaps, was that many tourists we spoke to had very little idea of where they actually were. The placelessness experienced by our tourists (*where are we?*), although in many ways radically different from the 'fictional' ones discussed by Elliott (*just another place*), is in part enabled by processes of enclaving: the creation of a 'tourist bubble' (Cohen, 1979; Judd, 1999) sealed off from the place around it. Enclaving extends 'backwards' to the pre-trip information – English-language websites, using an agent, all-inclusive trips and so on. But some experiences are much less enclaved than others. Global elites may stay in the best resorts, hotels and hospitals, and their money may insulate them from the surrounding culture, but our tourists' experience was frequently disruptive – sense of place came crashing rudely back in, in the form of Libyans with bullet wounds, locals sorting through rubbish tips behind the hospital and (imagined) dogs chasing rats in the night in Tunisia, for instance. As Edensor explains, even within the most mundane forms of tourism, 'meeting less contained forms of difference may involve a reflexive re-examination of numerous aspects of the everyday performance of tourism' (2007, p. 212). Suddenly our patients were connected not just to the local but to a set of *global* events widely reported on news programmes they did not watch. For some patients this profoundly impacted their consciousness; for others the enclave was not a physical barrier but rather a mental one that insulated them from this disruptive world of which they momentarily became a part (Mainil et al., 2010). Perhaps empathy is, after all, the privilege of the elite (Pedwell, 2012), and young women who pin their hopes on a little 'nip and tuck' in an unfamiliar location may have enough troubles of their own to worry about. Apparent enclaving can, then, mask more complex interactions between locals and tourists.

Of course, it is not just tourists who travel; doctors and nurses, medical techniques and technologies, ideas about beauty – these are all also on the move, and the interactions of these diverse flows are both predictable and disjunctive (Appardurai, 1990; Mainil et al., 2010); in some cases flows simply follow the money, but in others there are more complicated push-and-pull factors at work. The vagaries of currency rates are key, but so too is the tangibility of place and distance in determining the kinds of 'tourisms' that patients seek out – a luxurious family holiday or solitary flit to an operating theatre and hospital/hotel room. Place marks skill, hygiene, regulation, standards, usually by reference to practices in 'developed' countries; but it can also mark skill and technological development as historically indigenous, a mark of modernity and a source of national pride. It can also find medical cultures in even the most 'developed' of nations wanting by comparison.

Of course, the argument that it is wrong to call cosmetic surgery tourism *tourism*, because to do so trivialises the actual medical procedures, the risks and the pain, depends on a view that holidays are themselves trivial, rather than an important, even central, part of contemporary life. Nevertheless, it remains contentious among some commentators to use the label 'tourism'. From our perspective, this in itself raises interesting conceptual questions about what tourism means or can mean. We note the global importance of tourism as an industry, or rather as 'a hybrid economic formation blending different industries, the state, "nature", the informal sector, the capitalist and non-capitalist economies, and all manner of technologies, commodities and infrastructures' (Gibson, 2009, p. 529). Far from trivialising, we argue that using the moniker cosmetic surgery tourism reminds us to keep all these things in view. Tourism literature often foregrounds the accidental transformation of the tourist self in the face of disruptions to the planned holiday. But what does it mean when the transformation of the self is the exact aim of the holiday? And what if that transformation of the self is arrived at through a transformation of the body? These questions remain to be answered as we progress with our research.

Our aim in this chapter has been to explore cosmetic surgery tourism as a phenomenon that assembles a complex set of people, places and practices – cosmetic surgery tourism as temporary assemblage. We have explored how tourist destinations parlay particular ideas, images, myths and stories about place in order to attract visitors, including would-be consumers of cosmetic surgery. We looked at the ways in which popular cosmetic surgery tourism destinations are produced by existing tourist knowledge, the availability of cheap flights and favourable exchange rates, and by essentialist ideas about their citizens, particularly in relation to their caring, amenable or hospitable 'natures'. Finally, we considered debates in tourism studies to understand what it means to call our subject cosmetic surgery *tourism*.

Acknowledgements

We would like to thank the ESRC (Grant Reference RES-062-23-2796). Thanks also go to the editors of this volume for helpful comments on an earlier draft.

Notes

1. www.somniomedical.com
2. http://www.somniomedical.com/cosmeticholiday_whychoosesomnio.html
3. www.mybodyandspirit.com.au

References

Ackerman, S. (2010) 'Plastic Paradise: Transforming Bodies and Selves in Costa Rica's Cosmetic Surgery Tourism Industry' *Medical Anthropology* 29, 4, 403–23.

Augé, M. (1995) *Non-places: Introduction to an Anthropology of Hypermodernity* (London: Verso).

Bell, D., Holliday, R., Jones, M., Probyn, E. and Sanchez Taylor, J. (2011) 'Bikinis and Bandages: An Itinerary for Cosmetic Surgery Tourism' *Tourist Studies* 11, 2, 137–53.

Birch, J., Caulfield, R. and Ramakrishnan, V. (2007) 'The Complications of "Cosmetic Tourism" – An Avoidable Burden on the NHS' *Journal of Plastic, Reconstructive and Aesthetic Surgery* 60, 9, 1075–7.

Burkett, L. (2007) 'Medical Tourism: Concerns, Benefits, and the American Legal Perspective' *Journal of Legal Medicine* 28, 2, 223–45.

Casanova, E. (2007) 'The Whole Package: Exploring Cosmetic Surgery Tourism', Paper presented at the American Sociological Association Annual Conference, New York.

Castonguay, G. and Brown, A. (1993) ' "Plastic Surgery Tourism" Providing a Boon for Costa Rica's Surgeons' *Canadian Medical Association Journal* 148, 1, 74–6.

Cockerham, G. and Cockerham, W. (2012) *Health and Globalization* (London: Sage).

Cohen, E. (1979) 'A Phenomenology of Tourist Experiences' *Journal of the British Sociological Association* 13, 2, 179–201.

Connell, J. (2006) 'Medical Tourism: Sea, Sun, Sand and... Surgery' *Tourism Management* 27, 1093–100.

Connell, J. (2011) *Health and Medical Tourism* (Oxford: CABI).

Edensor, T. (2001) 'Performing Tourism, Staging Tourism: (Re)producing Tourist Space and Practices of Mundane Tourism' *Tourist Studies* 1, 1, 59–81.

Edensor, T. (2007) 'Mundane Mobilities, Performances and Spaces of Tourism' *Social and Cultural Geography* 8, 2, 199–215.

Edmonds, A. (2007) ' "The Poor Have the Right to Be Beautiful": Cosmetic Surgery in Neoliberal Brazil' *Journal of the Royal Anthropological Society* 13, 363–81.

Edmonds, A. (2010) *Pretty Modern: Beauty, Sex and Plastic Surgery in Brazil* (Durham: Duke University Press).
Elliott, A. (2008) *Making the Cut: How Cosmetic Surgery Is Transforming Our Lives* (London: Reaktion).
Gibson, C. (2009) 'Geographies of Tourism: Critical Research on Capitalism and Local Livelihoods' *Progress in Human Geography* 33, 4, 527–34.
Gimlin, D. (2007) 'Accounting for Cosmetic Surgery in the USA and Great Britain: A Cross-cultural Analysis of Women's Narratives' *Body and Society* 13, 1, 46–60.
Glinos, I., Baeten, R., Helble, M. and Maarse, H. (2010) 'A Typology of Cross-border Patient Mobility' *Health and Place* 16, 1145–55.
Goodrich, J. and Goodrich, G. (1987) 'Health-care Tourism – An Exploratory Study' *Tourism Management* September, 217–22.
Holliday, R., Hardy, K, Bell, D., Jones, M., Probyn, E. and Sanchez Taylor, J. (2013) 'Beautiful Face, Beautiful Place: Relational Geographies and Gender in Cosmetic Surgery Tourism Websites' *Gender, Place and Culture.*
Iyer, P. (1997) 'The Nowhere Man' *Prospect Magazine* http://www.prospectmagazine.co.uk/1997/02/thenowhereman/ date accessed 10 September 2012.
Jones, M. (2008) *Skintight: An Anatomy of Cosmetic Surgery* (Oxford: Berg).
Jones, M. (2011) 'Clinics of Oblivion: Makeover Culture and Cosmetic Surgery' *Portal: Journal of Multidisciplinary International Studies* 8, 2.
Judd, D. (1999) 'Constructing the Tourist Bubble' in S. Fainstein and D. Judd (eds) *The Tourist City* (Tale: Yale University Press).
Kangas, B. (2010) 'Traveling for Medical Care in a Global World' *Medical Anthropology* 29, 1, 311–62.
Lunt, N., Hardey, M. and Mannion, R. (2010) 'Nip, Tuck and Click: Medical Tourism and the Emergence of Web-based Health Information' *The Open Medical Informatics Journal*, 4, 1–11.
Mainil, T., Platenkamp, V. and Meulemans, H. (2010) 'Diving into the Contexts of In-between Worlds: Worldmaking in Medical Tourism' *Tourism Analysis*, 15, 743–54.
Mainil, T., Platenkamp, V. and Meulemans, H. (2011) 'The Discourse of Medical Tourism in the Media' *Tourism Review*, 66, 1, 31–44.
Pedwell, C. (2012) 'Affective (Self-) Transformations: Empathy, Neoliberalism and International Development' *Feminist Theory*, 13, 2, 163–179.
Ramirez de Arellano, A. (2007) 'Patients without Borders: The Emergence of Medical Tourism' *International Journal of Health Services* 37, 1, 193–8.
Reisman, D. (2010) *Health Tourism: Social Welfare through International Trade* (Cheltenham: Edward Elgar).
Rojek, C. (1997) 'Indexing, Dragging and the Social Construction of Tourism Sights' in C. Rojekand J. Urry, (eds) *Touring Cultures: Transformations of Travel and Theory* (London: Routledge).
Schult, J. (2006) *Beauty from Afar* (New York: Stewart, Tabori and Chang).
Urry, J. (1990) *The Tourist Gaze* (London: Sage).
Urry, J. (1995) *Consuming Places* (London: Routledge).

7

Cross-Border Reproductive Care Around the World: Recent Controversies

Wannes Van Hoof and Guido Pennings

The key arguments of this chapter are as follows:

- Cross-border reproductive care (CBRC) is a worldwide phenomenon, but there is a lack of empirical data on patterns of movement and on the experiences of reproductive travellers. Several push and pull factors have been identified in preliminary data.
- Assisted reproductive technologies are often culturally, religiously or ethically controversial. Both the technologies themselves and the convictions about them are evolving rapidly. Cross-border reproductive care adds to this complexity.
- The ethical differences have led to legal diversity across the world. Patients are travelling from restrictive states to permissive states to make use of assisted reproduction. This leads to questions about law evasion, tolerance and the validity of certain restrictions.
- Legal diversity has problematic consequences for gamete donation across borders or international commercial surrogacy. The laws of two countries are often not equipped to regulate the birth of a child from cross-border reproductive care. The issues of compensation, identifiability and exploitation all pose additional ethical challenges in cross-border situations.
- It will be necessary to continue ethical reflection on cross-border reproductive care as the phenomenon develops further.

Introduction

Cross-border reproductive care (CBRC) is a growing phenomenon where infertile patients cross borders in order to obtain reproductive treatment abroad. CBRC has many names. Twenty years ago, Knoppers and LeBris (1991) were the first to describe what they called 'procreative tourism'.

In their wake, the phenomenon was also called 'fertility tourism' and 'reproductive tourism', which seems logical when CBRC is considered a subset of medical tourism. However, medically assisted reproduction is a very specific form of medical treatment with different ethical, legal, social and cultural consequences. Tourism implies travelling for pleasure, and by no means do infertile couples travel for pleasure. Travelling for reproductive care may be challenging, time-consuming, frustrating, impoverishing and frightening (Inhorn and Patrizio, 2009). 'Reproductive exile' was suggested as an alternative to the tourism discourse since exile means leaving one's country, usually for political reasons (Matorras, 2005). The current concept of cross-border reproductive care avoids the connotations of pleasure and leisure time implied in the term 'tourism'; it is objective and descriptive since it holds no value judgement regarding the movements and it links with the more general term 'cross-border health care' that is commonly used when other types of movements for health services are considered (Pennings, 2005).

The different causes of CBRC can be divided into two groups: legal restrictions and/or availability issues in the state of affiliation (Pennings et al., 2008). A very important aspect of CBRC that distinguishes it from medical tourism is the issue of legal diversity (Van Hoof and Pennings, 2012). Many fertility treatments (for example, egg donation, commercial surrogacy) or techniques (for example, sex selection) are prohibited in one state and actively advertised in another. Some groups of people (for example, same-sex couples, single parents, women over a certain age) who are denied access to medically assisted reproduction in one country are treated without any issues in another. The consequence of this diversity is a flow of would-be parents from states with restrictive legislation to more permissive countries (Storrow, 2005). This form of cross-border care has been categorised as 'circumvention tourism' or even 'rogue medical tourism' (Hunter and Oultram, 2010; Cohen, 2012). However, it may be more accurate to see it as a form of moral pluralism in motion (Pennings, 2002).

Whether patients travel because of legal prohibitions or because the treatment they need is not available at home (within a reasonable timeframe), most are disappointed with the fertility sector at home. They feel they are being denied the basic right to have children. For example, the UK has a very liberal policy on medically assisted reproduction, but a shortage of donor gametes. As a consequence, the most important motivation for UK patients to travel abroad is the desire for timely and affordable treatment with donor gametes (Culley et al., 2011). These patients feel that they have been let down by the state, whose legislation on gamete donation should be more attractive for donors. However, the state may have valid moral reasons for regulating gamete donation with limited compensation fees and with identifiability of the donor. The tension between the individual and the state will reappear throughout this chapter.

We will present an overview of CBRC, showing that it is a worldwide phenomenon and specifying some underlying patterns. Subsequently we will turn to some of the ethical controversies in CBRC. We will address the issue of legal diversity, examine the ethics of gamete donation in CBRC and identify some problems with international commercial surrogacy in the context of CBRC.

Cross-border reproductive care around the world

There is a general lack of data on the prevalence of CBRC. Our focus will mainly be on Europe, because that is where the most reliable data come from at the moment and because the issue of legal diversity is very apparent there. Much of the data that have been published is based on very gross estimates and the studies are often methodologically lacking (for example, Nygren et al., 2010). In a recent survey of patients in 44 clinics in 6 countries it was estimated that at least 12,000–15,000 patients cross borders for fertility treatment every year within Europe (Shenfield et al., 2010). No reliable data are available for North America, but physicians estimate it to be a widespread phenomenon (Hughes and DeJean, 2010). In Asia, several countries have specialised clinics that are frequented by foreigners for specific treatments (for example, commercial surrogacy) or applications (for example, sex selection) (Pande, 2009; Whittaker, 2011a). In South America, the private health-care sector has discovered the market for good fertility care in Argentina, Brazil, Chile and Mexico (Smith et al., 2010). In the Middle East, Muslim pronatalist culture is starting to result in a thriving assisted reproductive technology (ART) industry, but locally there still are many structural and cultural constraints and there is uncertainty about the ethics of gamete donation in Islam. It is forbidden in Sunni Islam but allowed in Shia Islam. The use of donor gametes is considered to be controversial by many Muslims, Sunni or Shia. Therefore, those Muslim couples who want to make use of third-party assisted reproduction generally cross borders because of privacy and socio-ethical reasons, often to Iran, where Shia Islam is dominant (Abbasi-Shavazi et al., 2008).

Patterns of movement: Push and pull

CBRC is a growing phenomenon. Several factors are responsible for this growth. Worldwide, at least 15 per cent of reproductive-age couples suffer from infertility (Inhorn, 2009). There are many potential patients who are willing and able to pay and travel for excellent care. This is made possible by globalisation, which makes transnational transfer of people, technologies and ideas ever easier, and by commercialisation of assisted reproduction (Gürtin and Inhorn, 2011).

Crossing borders for health care has become easier. Within Europe, for example, the recent directive on the application of patients' rights in cross-border health care (2011/24/EU) will give patients the right to be reimbursed for care in any EU Member State without prior authorisation up to the limit to which the patient is entitled according to the legislation of the home country. This may prove to be a great instrument to help patients who face long waiting lists.

Traditionally, patterns of movement in CBRC are described only by referring to the push factors. These include a type of treatment forbidden by law for moral reasons; a treatment not available because of lack of expertise or equipment or for safety reasons; certain categories of patients who are not eligible for assisted reproduction; waiting lists (for example, because of a lack of gamete donors); and costs (Pennings, 2002). The focus on push factors stems from the fact that patients generally need to be pushed abroad: they do not want to cross borders for treatment, especially if it is as emotional and personal as fertility treatment.

However, that is only half the story. When patients decide to travel, they still have to decide where to travel to. To that end, patients conduct endless Google searches, read blogs, join support groups and find specialised brokers (Speier, 2011). Because reproduction is such a private matter, patients are pulled towards countries they have some kind of connection with. For example, recent research in an Iranian clinic indicates that Muslim patients want to be treated in a religiously correct fashion (Moghimehfar and Nasr-Esfahani, 2011).

A minority of patients choose to go abroad for fertility treatment even when they could be helped at home. These patients are typically expatriates who return home for medically assisted reproduction because of medical patriotism, a shared language with the medical personnel, co-religion, moral trustworthiness, donor phenotype, the comforts of home and discrimination (Inhorn, 2011). Several traditional emigration regions, like Thailand and the Middle East, are now destination countries for reproductive services to their expatriates (Whittaker, 2009; Inhorn, 2011). In this case, patients are pulled towards a specific country.

Most patients who are planning treatment abroad are looking for a solution to the problem that is pushing them away from home. Several destination countries have an interest in attracting as many of these reproductive travellers as possible. They are interested in the promise of financial gains by treating rich foreign patients. To accommodate these patients, they are looking to pull them in with lower costs, high quality of care, timely treatment and favourable legislation. For example, North American patients are faced with Canadian donor shortages and high costs in the US and become 'consumers shopping for discount assisted reproduction treatments': brokers advertise treatments abroad that ensure similar donor phenotypes and high egg donor availability at competitive prices (Speier, 2011). One of the main pull factors in CBRC is the availability of a group of women ready to act as surrogates or as egg donors. The lack of such women is the main reason for waiting lists in Canada and European countries and for the high costs in the US. In developing economies it is easier to find available women, but this increases the risk of exploitation given the unequal economic position (Whittaker, 2011b). For some controversial treatments, the mere offer of a treatment may be a pull factor. For example, a Thai clinic suggested that the availability of social sex selection was a prime motivator for travelling among its Vietnamese, Indian and Chinese patients (Whittaker, 2011a).

Controversies in cross-border reproductive care

Medically assisted reproduction is rapidly changing. In 1991, Knoppers and LeBris wrote that there were several areas of consensus: access to fertilisation techniques should be limited to heterosexual married couples or to those living in stable unions, commercial surrogacy agencies or intermediaries should be prohibited, and reproductive technologies should be free from commercialisation. This is definitely no longer the case. There are many controversies surrounding medically assisted reproduction because human reproduction is subject to very strong moral beliefs that differ greatly within and between societies. Is there a right to have children? Does this right extend to everyone (for example, same-sex couples, single parents, infertile couples)? Should societies support procreation (for example, public funding)? Moreover, medically assisted reproduction introduces treatments that are controversial themselves. For example, most treatments imply loss of embryos, other treatments depend on third parties who may risk harm or exploitation (for example, egg donors, surrogates). In general, assisted reproduction challenges traditional (natural) norms of conception and parenthood.

In addition to dynamics at national levels concerning assisted reproduction, ever-changing flows of patients add to the complexity of CBRC. In the next sections, we will focus on the ways in which crossing borders causes problems or controversies. First, we will address some of the general issues that arise from the lack of general consensus: what are the consequences of having this legal diversity? Subsequently, we will turn to gamete donation and international commercial surrogacy to see what legal diversity means for concrete cross-border treatments. Because of the complexity of the situations we address, it is impossible to suggest a straightforward solution for many problems. However, it seems that a more permissive regulation within a country could prevent many problems. One a priori insight could prevent many problems: when a treatment or technique is allowed in the country, people will no longer be forced to go abroad. Moreover, it would enable the government to regulate the technique in a way that is morally acceptable and to devise safeguards against abuse.

Legal diversity

Some countries prohibit some forms of medically assisted reproduction (for example, gamete donation) or deny access to some groups of people (for example, same-sex couples) while other countries permit the same actions. Legal diversity is a necessary condition for cross-border movements for law evasion. Several patterns have developed over the years, directly affected by legal developments. Many European countries in close geographical proximity have very different regulations regarding assisted reproduction. The permissive policies on egg donation in Spain and Czech Republic have made them popular destinations for couples who require donor eggs (Bergmann, 2011). Many Italians go to Spain and many Germans to Czech Republic to evade restrictions on egg donation in their home state (Shenfield et al., 2010). Dutch clinics are reluctant to accept women over 40, so these women turn to Belgian clinics that are open to treatment of older patients (Pennings et al., 2009). In France, only heterosexual couples have access to fertility treatments, so many French lesbians travel to Belgium to be inseminated with donor sperm (Pennings et al., 2009). In Italy, reproductive emigration quadrupled when the restrictive law of 2004 was enacted. This law prohibits gamete donation and surrogacy entirely and restricts access to medically assisted reproduction to heterosexual couples (Boggio, 2005; Ferraretti et al., 2010). Italians who cannot conceive in Italy because of this law feel abandoned by their state (Zanini, 2011).

Legal diversity has been a major point of discussion since CBRC became a prominent subject in academics (Pennings, 2002). One central question is how CBRC for law evasion affects national legislation. It was first argued that CBRC acts as a 'safety valve' for the moral minorities in society, as a way for them to circumvent the majoritarian restrictions on assisted reproduction (Pennings, 2004). Following this view, a country should allow CBRC for law evasion as a form of tolerance. The movements by patients to other countries could change current legislation if the lawmakers would be aware of the extent of the problem. If the lawmakers were to realise that patients who need certain restricted treatments go abroad, they might soften the restrictions. This can only be the case if the movements are visible to the larger public, but this condition is difficult to fulfil as patients generally do not want to go public with their highly personal and private problems (Pennings et al., 2008). All in all, the effect of a safety valve is primarily that it decreases pressure for law reform internally. The opportunity for patients to go abroad can temper organised resistance to the law and allow governments to pass more restrictive legislation than it otherwise might (Storrow, 2010). This means legal diversity may inherently lead to more restrictions within countries.

Recent developments have shown that legal diversity can have a direct effect on national legislation: in extreme cases the growing phenomenon of

CBRC for law evasion has led to a total ban on reproductive travelling. This happens when a radical moral belief, often religiously inspired, is the basis for legislation. In that case, no tolerance in the form of CBRC can be allowed. In Turkey and several states in Australia, respectively, gamete donation and commercial surrogacy were made punishable offences, also when performed outside the national territory (Van Hoof and Pennings, 2011). In Turkey, for example, CBRC was very visible: several clinics had sophisticated arrangements with Cypriote clinics to circumvent the restrictive policy (Gürtin, 2011). The effect of this visibility was not that the lawmakers softened the restrictions. On the contrary, they maintained the restrictions and tried to block the escape routes. The movements did not improve the situation for local patients. Moreover, the possibility of CBRC may be used as an excuse to defend the status quo. For example, in a recent case the European Court of Human Rights upheld Austria's restrictions on gamete donation, partly because Austrian patients always have the option of going abroad (Van Hoof and Pennings, 2012). This has recently been called the pluralism problem in CBRC (Storrow, 2010).

While this may be the main problem in the grand scheme of things, for the current travelling patients legal diversity may cause a range of challenging problems as well. If there is a criminal law against a procreation technique, as is the case in Turkey and Australia, patients may face charges when they return. In the next two sections, we will look at what practical problems legal diversity can cause for patients who undergo gamete donation abroad or engage in international commercial surrogacy.

Gamete donation

There are many controversies surrounding gamete donation in general. These moral complexities become even more difficult in a cross-border situation. Here, we will focus on how crossing borders complicates the issues of compensation/payment and donor identifiability.

Compensation/payment: Balancing justice and autonomy

In the discussion on gamete donation, it is argued that there is a significant moral difference between compensating someone for their time and inconvenience, and paying someone for their eggs or sperm. In this line of argument, justice requires that a donor is compensated. On the other end of the discussion the argument is made that gamete donors, especially egg donors, are selling their body parts (Almeling, 2007). This is clearly the case when donors are paid per egg they produce at the end of their stimulated cycle. In the same line of argument, some even compare travelling for eggs with transplant tourism (Pfeffer, 2011). It is true that in some instances of abuse, egg donors are treated badly and over-stimulated during the donation process and compensated only very little (Nahman, 2011). However, egg selling and overstimulation are malpractices that can and should be prevented: potential harm is not a sufficient reason to ban a practice as a whole. There are guidelines for physicians to ensure ethical practice of CBRC (Shenfield et al., 2011).

The issue then is to decide what counts as fair compensation and what constitutes payment. Those who oppose compensation for gamete donation argue that payment of big sums often leads to coercion and undue inducement. This implies that the autonomy of the donor is not respected. However, minimal compensation for the time spent and inconvenience suffered for egg donation is equally immoral, especially when the infertile woman pays a large sum for her treatment and the clinics and physicians make huge profits using those eggs. In this case, the balance between time and effort on the one hand and compensation on the other is skewed.

In practice, there is a big difference in the compensation that donors receive across the world. In the UK, egg donors are compensated £750 (this used to be £250 plus expenses), in Czech Republic €800 and in Spain €900. One might say that a European consensus is slowly emerging. However, in Romania, egg donors are paid $200, which arguably constitutes unfair or inadequate compensation. In the US, there is no cap on payments to donors. There are only guidelines: the American Society for Reproductive Medicine recommends a cap of $5,000 and called sums over $10,000 inappropriate (American Society for Reproductive Medicine, 2007). However, offers of $50,000 have been reported to 'an extraordinary egg donor' (Levine, 2010). Clearly, this no longer constitutes mere compensation. If you are looking for an instance of undue inducement, this would be it: that kind of money is a strong incentive for anyone to donate.

The balance between the basic principles of justice and autonomy is complicated by cross-border issues. A reasonably capped compensation is still no guarantee that prospective donors are not unduly induced. Foreign women, for whom the same amount of money can have a totally different meaning, can be attracted by the compensation fee. The availability of donors because of high compensation fees may attract a large population of foreign patients, which may in turn increase the demand for egg donors which may lead to clinics giving them more incentives to donate repeatedly.

The balance between justice and autonomy is even more precarious when frozen eggs cross borders. Freezing technology allows for an invisible current of gametes crossing borders instead of donors. Donor sperm is already being shipped across borders for years and recent developments in egg freezing will allow the same for donor eggs. The question is whether these overseas donors should be compensated the same amount as local donors, which may constitute a larger sum in their place of origin.

In Ireland, there is already a clinic that transfers frozen sperm from the prospective father to Ukraine, where the eggs of a donor are fertilised (Walsh et al., 2010). The resulting embryos are frozen and sent back to

Ireland, where they are implanted. Should the Ukrainian donors be compensated the same amount as Irish donors? Should they receive the same amount as other Ukrainian donors? To protect the autonomy of the egg donors, they should be compensated only the same amount as other Ukrainian donors. This would be in accordance with the principle of justice: compensation for time and effort must be related to purchasing power. However, it is unjust to let the clinic absorb the difference in purchasing power as profit. It is equally unjust to allow a two-tier system of domestic and foreign eggs at different prices. This would result in a very commercialised form of egg donation. The reproductive labour of egg donors should not be seen as something that can be exported to countries with lower wages. This cross-border problem is a very complex ethical issue that will demand more reflection as the situation progresses.

Donor identifiability

Both in the academic literature and in legislation across the world, a trend towards more identifiability can be identified (Blyth and Farrand, 2004). Some believe the removal of donor anonymity will deter people from donating their gametes. This is contested by others, who argue that the typical donor will no longer be a student, but someone older, often a parent, and that different recruitment strategies are needed when anonymity is removed (Frith et al., 2007).

Removal of donor anonymity is reported to be a reason to travel abroad for gamete donation. For example, Dutch patients travel to Belgium for anonymous donation (Pennings et al., 2009). Twenty-six per cent of UK women indicated 'a wish for anonymous sperm donation' as their main reason to travel abroad (Shenfield et al., 2010). Some fertility travellers state that donor anonymity is a by-product of their journey because it is mandatory where they were treated; it is not their main reason for travelling (Culley et al., 2011).

When patients who live in a country with identifiable donation make use of anonymous donation abroad, they act against a societal trend. The interest in genetic origins has become a hype in recent years. However, when patients cross borders from a country where gamete donors are identifiable to a country where donation is anonymous (for example, because of a donor shortage in the home country), their child will grow up in a society where all other donor-conceived children can in principle find out about their genetic origins. If the parents are open about the donor conception, the children face the extra burden of a foreign anonymous donor in the age of donor-sibling registries and TV shows titled 'Who is my father?'

'Identifiability' in fact means 'contactability': whether or not your donor is identifiable may not really matter if he does not speak your language, comes from a very different culture and lives far away in another country. As far as being able to contact your genetic parent is concerned, it may not even matter whether patients make use of an anonymous or an identifiable donor abroad: they are all hardly contactable. Cryobanks are already shipping sperm from 'identifiable' donors across borders to countries that enforce identifiable donation. Due to persisting shortages this practice is likely to increase, but while it is technically within the legal requirements for imported sperm, it is incompatible with the spirit of the law.

International commercial surrogacy

The issue of compensation and payment is also present in discussions on commercial surrogacy, but it runs largely parallel with the discussions on gamete donation. Here, we will focus on the exploitation argument and on issues of comity. Issues of comity are problems commissioning couples face when the legal system of the country where they were treated is not 'compatible' with their domestic laws (Storrow, 2011).

Exploitation

It is not easy to define exploitation. The exploitation argument has been used against commercial surrogacy in many different ways: sometimes it means commodification of the body, sometimes undue inducement, sometimes taking advantage of vulnerable women. In the case of international commercial surrogacy, it is a fact that surrogates often 'work' in abysmal conditions and that there is a huge disparity between the power and wealth of the commissioning parents and that of the surrogate. The first step in making commercial surrogacy less 'exploitative' would be to improve the conditions under which the surrogates work and to change the background inequalities which generate exploitative relationships (Wilkinson, 2003).

Feminist authors have dismissed such arguments as secondary to the real problem. Using a woman as some kind of breeder, a womb for rent, is intrinsically wrong regardless of context or consequences (Ber, 2000). Ultimately, this is based on the Kantian categorical imperative never to use a human being merely as a means to an end, but always as a person. This ethical objection can be overturned if the practice of commercial surrogacy is possible while respecting the surrogate also as a person. Parks (2010) has suggested that the relationship between the commissioning couple and the surrogate should be governed by care and interpersonal relationships rather than by commercial interests, in which case international commercial surrogacy is not exploitative. Recent ethnographic fieldwork suggests that both the commissioning couple and the surrogate resist the commercial and contractual nature of their relationship by establishing some kind of 'sisterhood' with each other (Pande, 2011).

Even if the nature of the relationship with the surrogate is not personal, she can be respected as a person. If we conceptualise surrogacy as reproductive labour, commercial surrogacy is not exploitative as long as there are fair compensations, good regulations, good working conditions and medical support. In the case of international commercial surrogacy, one might say there that there should be 'fair trade international surrogacy' (Humbyrd, 2009). In that case, a woman can autonomously decide to partake in a valuable surrogacy project.

Issues of comity

Many surrogacy contracts in frequently visited countries for international commercial surrogacy, such as India, are not legally binding because there is no legislation on the matter (yet). However, this does not deter patients who need surrogacy from engaging in the practice. This leads to a lot of uncertainty for both parties in the arrangement: the commissioning couple could be extorted for more money after the child is born and the surrogate mother could decide to keep the child. The surrogate mother could be left with the child, especially in case of birth defects; she could be denied necessary medical care after she gave birth. However, in practice these problems seem to occur very rarely (Busby and Vun, 2010).

There may be few problems between the couple, the surrogate and the clinic; problems may still arise at state level. Even when the visited country has appropriate legislation on the matter and there is an enforceable contract in place, issues of comity may arise after the birth of the child. For example, when a same-sex couple is legally married at home and they move abroad, it is an issue of comity whether or not their marital bond will be recognised. When same-sex couples who are denied a right to marry or to adopt a child go abroad to conceive a child, the partner who does not share a genetic link to the child cannot legally recognise it. A similar problem arises in the case of surrogacy contracts across borders: the contract between the commissioning couple and the clinic/surrogate may be considered valid in the country of treatment, but it will not be recognised in the home country of the couple if commercial surrogacy is illegal or unregulated there. When a surrogate has delivered the baby, the surrogacy contract determines who the parents of the child are. Usually, the visited country considers the commissioning couple to be the legal parents. However, when the contract is not recognised by their home country, the woman who gave birth, the surrogate, is considered to be the legal parent. In practice, this leaves these children in a legal limbo, trapped between two legal systems, stateless, legally parentless and without passports to go home. Such problems have been reported for parents from the UK who went to Ukraine (Theis et al., 2009); Belgian parents who went to the US and to Ukraine (Van Hoof and Pennings, 2012) ; Japanese parents who went to India (de Alcantara, 2010; Parks, 2010) ; French parents who went abroad, mostly to the US (Rotman, 2009); German parents who went to India (Mahapatra, 2010); and many others. These cases generally end in court, where a judge allows an exception to the law 'in the best interest of the child'. We will address the situation in Ireland to illustrate that it is important to have some kind of regulation and the situation in Germany to show that even with clear prohibitive legislation problems of legal diversity can persist.

Currently, there is a lot of media attention for 15 Irish children who are 'stuck' in India (O'Brien, 2011a). Their parents are tired and angry with the Irish authorities. First, the lack of regulation in Ireland forced them to travel abroad. Subsequently, the lack of regulation prevents them from returning with their child. One couple is planning to take their case to the European Court of Human Rights because their right to private and family life is being infringed due to the lack of specific legal provisions which allow them to become their child's parents (O'Brien, 2011b). They would have a good case if the right to private and family life entails a right to assisted reproductive technologies. However, previous rulings indicate that the court does not deem this to be the case.

In Germany, commercial surrogacy is explicitly banned. When German couples go abroad for commercial surrogacy, they evade the domestic law and commit an immoral act from the point of view of their restrictive country. When they want to bring their child home, they are actually asking the state to directly contradict its policy by tolerating their law evasion. If the state takes its own moral position seriously, it should be consistent and deny the claim to parental rights. In this case the state cannot be found lacking as long as the laws are morally justified. The judges are caught between the devil and the deep blue sea: if they uphold the law, the child (and family) suffers; if they give in, more couples will go abroad to do the same.

In the current context of legal diversity, the strong wish for a child drives patients to controversial decisions. When they travelled abroad these parents were either uninformed/badly informed or they consciously put themselves and their child through a long and stressful legal battle, during which their child may be unable to return to the country of its parents and has to be cared for in difficult circumstances. Until now, the courts were always persuaded by the arguments that refer to the welfare of the child and allowed exception after exception. We have not seen the last of international commercial surrogacy and the issue of comity.

Conclusion

There is an urgent need for more reliable data on patterns of CBRC worldwide. This will inform policy makers of the exodus or influx of patients in their country and will allow them to react appropriately. Moreover, the patterns of movement will indicate changing attitudes and increasing needs among the population. More data should also be gathered on the experience of reproductive travellers. This will help ethical reflection on CBRC and will inform clinics and physicians of the needs of their international patients.

Many assisted reproductive technologies are ethically controversial. They are subject to rapidly evolving scientific research and changing views in public morality. Cross-border situations introduce new difficulties and complexities to assisted reproduction. In addition, cross-border flows are subject to dynamic conditions like legislation, availability of gamete donors and surrogates, and quality of care. It will be necessary to continue ethical reflection on CBRC as the phenomenon develops further.

Bibliography

Abbasi-Shavazi, M. J., Inhorn, M. C., Razeghi-Nasrabat, H. B. and Toloo, G. (2008) 'The "Iranian ART revolution": Infertility, assisted reproductive technology, and third-party donation in the Islamic Republic of Iran' *Journal of Middle East Women's Studies* 4, 2, 1–28.

Almeling, R. (2007) 'Selling genes, selling gender: Egg agencies, sperm banks, and the medical market in genetic material' *American Sociological Review* 72, 3, 319–40.

American Society for Reproductive Medicine (2007) 'Financial compensation of oocyte donors' *Fertility and Sterility* 88, 2, 305–9.

Ber, R. (2000) 'Ethical issues in gestational surrogacy' *Theoretical Medicine and Bioethics* 21, 153–60.

Bergmann, S. (2011) 'Reproductive agency and projects: Germans searching for egg donation in Spain and the Czech Republic' *Reproductive Biomedicine Online* 23, 600–8.

Blyth, E. and Farrand, A. (2004) 'Anonymity in donor-assisted conception and the UN Convention on the rights of the child' *International Journal of Children's Rights* 12, 89–104.

Boggio, A. (2005) 'Italy enacts new law on medically assisted reproduction' *Human Reproduction* 20, 5, 1153–7.

Busby, K. and Vun, D. (2010) 'Revisiting the handmaid's tale: Feminist theory meets empirical research on surrogate mothers' *Canadian Journal of Family Law* 26, 13–93.

Cohen, G. I. (2012) 'Circumvention tourism' *Cornell Law Review* 97, 1–92.

Culley, L., Hudson, N., Rapport, F., Blyth, E., Norton, W. and Pacey, A. A. (2011) 'Crossing borders for fertility treatment: Motivations, destinations and outcomes of UK fertility travellers' *Human Reproduction* 26, 9, 2373–81.

de Alcantara, M. (2010) 'Surrogacy in Japan: Legal implications for parentage and citizenship' *Family Court Review* 48, 3, 417–30.

Ferraretti, A. P., Penning, G., Gianaroli, L., Natali, F. and Magli, M. C. (2010) 'Cross-border reproductive care: A phenomenon expressing the controversial aspects of reproductive technologies' *Reproductive Biomedicine Online* 20, 261–6.

Frith, L., Blyth, E. and Farrand, A. (2007) 'UK gamete donors' reflections on the removal of anonymity: Implications for recruitment' *Human Reproduction* 22, 6, 1675–80.

Gürtin, Z. (2011) 'Banning reproductive travel: Turkey's ART legislation and third-party assisted reproduction' *Reproductive Biomedicine Online* 23, 555–64.

Gürtin, Z. and Inhorn, M. C. (2011) 'Introduction: Travelling for conception and the global assisted reproduction market' *Reproductive Biomedicine Online* 23, 535–7.

Hughes, E. J. and DeJean, D. (2010) 'Cross-border fertility services in North America: A survey of Canadian and American providers' *Fertility and Sterility* 94, 1, e16–e19.

Humbyrd, C. (2009) 'Fair trade international surrogacy' *Bioethics* 9, 111–8.

Hunter, D. and Oultram, S. (2010) 'The ethical and policy implications of rogue medical tourism' *Global Social Policy* 10, 297–9.

Inhorn, M. C. (2009) 'Right to assisted reproductive technology: Overcoming infertility in low-resource countries' *International Journal of Gynecology and Obstetrics* 106, 172–4.

Inhorn, M. C. (2011) 'Diasporic dreaming: Return reproductive tourism to the Middle East' *Reproductive Biomedicine Online* 23, 582–91.

Inhorn, M. C. and Patrizio, P. (2009) 'Rethinking reproductive "tourism" as reproductive "exile"' *Fertility and Sterility* 92, 3, 904–6.

Knoppers, B. M. and LeBris, S. (1991) 'Recent advances in medically assisted conception: Legal, ethical and social issues' *American Journal of Law and Medicine* 17, 4, 329–61.

Levine, A. (2010) 'Self-regulation, compensation and the ethical recruitment of oocyte donors' *Hastings Center Report* 40, 2, 25–36.

Mahapatra, D. (2010) 'German surrogate twins to go home' *Times of India*, 27 May, http://articles.timesofindia.indiatimes.com/2010-05-27/india/28279835_1_stateless-citizens-balaz-surrogate-mother date accessed 23 January 2012.

Matorras, R. (2005) 'Reproductive exile versus reproductive tourism' *Human Reproduction* 20, 12, 3571.

Nygren, K., Adamson, D., Zegers-Hochschild, F. and de Mouzon, J. (2010) 'Cross-border fertility care – International Committee Monitoring Assisted Reproductive Technologies global survey. 2006 data and estimates' *Fertility and Sterility* 94, 1, e4–e10.

Moghimehfar, F. and Nasr-Esfahani, M. H. (2011) 'Decisive factors in medical tourism destination choice: A case study of Isfahan, Iran and fertility treatments' *Tourism Management* 32, 1431–4.

Nahman, M. (2011) 'Reverse traffic: Intersecting inequalities in human egg donation' *Reproductive Biomedicine Online* 23, 626–33.

O'Brien, C. (2011a) 'Surrogacy children caught in legal limbo' *The Irish Times*, 19 November http://www.irishtimes.com/newspaper/frontpage/2011/1119/1224307824130.html date accessed 23 January 2012.

O'Brien, C. (2011b) 'Surrogacy guidelines to be issued next month' *The Irish Times*, 23 November http://www.irishtimes.com/newspaper/ireland/2011/1123/1224308000862.html#. Ts0lTJ3ZKv8.facebook date accessed 23 January 2012.

Pande, A. (2009) 'Not an "Angel", not a "Whore": Surrogates as "Dirty" Workers in India' *Indian Journal of Gender Studies* 16, 2, 141–73.

Pande, A. (2011) 'Transnational commercial surrogacy in India: Gifts for global sisters?' *Reproductive Biomedicine Online* 23, 618–25.

Parks, J. A. (2010) 'Care ethics and the global practice of commercial surrogacy' *Bioethics* 24, 333–40.

Pennings, G. (2002) 'Reproductive tourism as moral pluralism in motion' *Journal of Medical Ethics* 28, 337–41.

Pennings, G. (2004) 'Legal harmonization and reproductive tourism in Europe' *Human Reproduction* 19, 12, 2689–94.

Pennings, G. (2005) 'Reply: Reproductive exile versus reproductive tourism' *Human Reproduction* 20, 12, 3571–2.

Pennings, G., Autin, C., Decleer, W., Delbaere, A., lbeke, L., Delvigne, A., De Neubourg, D., Devroey, P., Dhont, M., D'Hooghe, T., Gordts, S., Lejeune, B., Nijs, M., Pauwels, P., Perrad, B., Pirard, C. and Vandekerckhove, F. (2009) 'Cross-border reproductive care in Belgium' *Human Reproduction* 24, 12, 3108–18.

Pennings, G., de Wert, G., Shenfield, F., Cohen, J., Tarlatzis, B. and Devroey, P. (2008) 'ESHRE Task Force on Ethics and Law 15: Cross-border reproductive care' *Human Reproduction* 23, 10, 2182–4.

Pfeffer, N. (2011) 'Eggs-ploiting women: A critical feminist analysis of the different principles in transplant and fertility tourism' *Reproductive Biomedicine Online* 23, 634–41.

Rotman, C. (2009) 'Gestation pour autrui: les enfants fantômes de la République. Libération' 20 May, http://www.liberation.fr/societe/0101568271-gestation-pour-autrui-les-enfants-fantomes-de-la-republique date accessed 10 September 2012.

Shenfield, F., de Mouzon, J., Pennings, G., Ferraretti, A. P., Nyboe Andersen, A., de Wert, G. and Goossens, V. (2010) 'Cross border reproductive care in six European countries' *Human Reproduction* 25, 6, 1361–8.

Shenfield, F., Pennings, G., de Mouzon, J., Ferraretti, A. P., and Goossens, V. (2011) 'ESHRE's good practice guide for cross-border reproductive care for centers and practitioners' *Human Reproduction* 26, 7, 1625–7.

Smith, E., Behrmann, J., Martin, C. and Williams-Jones, B. (2010) 'Reproductive tourism in Argentina: Clinic accreditation and its implications for consumers, health professionals and policy makers' *Developing World Bioethics* 10, 2, 59–69.

Speier, A. (2011) 'Brokers, consumers and the internet: How North American consumers navigate their infertility journeys' *Reproductive Biomedicine Online* 23, 592–9.

Storrow, R. F. (2005) 'Quests for conception: Fertility tourists, globalization and feminist legal theory' *Hastings Law Journal* 57, 295–330.

Storrow, R. F. (2010) 'The pluralism problem in cross-border reproductive care' *Human Reproduction* 25, 12, 2939–43.

Storrow, R. F. (2011) 'Assisted reproduction on treacherous terrain: The legal hazards of cross-border reproductive travel' *Reproductive Biomedicine Online* 23, 538–45.

Theis, L., Gamble, N. and Ghevaert, L. (2009) 'Re X and Y: "A trek through a thorn forest"' *Family Law* (March), 239–43.

Van Hoof, W. and Pennings, G. (2011) 'Extraterritoriality for cross-border reproductive care: Should states act against citizens travelling abroad for illegal infertility treatment?' *Reproductive Biomedicine Online* 23, 546–54.

Van Hoof, W. and Pennings, G. (2012) 'Extraterritorial laws for cross-border reproductive care: The issue of legal diversity' *European Journal of Health Law* 19, 187–200.

Walsh, D., Omar, A. B., Collins, G. S., Murray, G. U., Walsh, D. J., Salma, U. and Sills, E. S. (2010) 'Application of EU Tissue and Cell Directive screening protocols to anonymous oocyte donors in western Ukraine: Data from an Irish IVF programme' *Journal of Obstetrics and Gynaecology* 30, 613–16.

Whittaker, A. (2009) 'Global technologies and transnational reproduction in Thailand' *Asian Studies Review* 33, 319–32.

Whittaker, A. (2011a) 'Reproduction opportunists in the new global sex trade: PGD and non-medical sex selection' *Reproductive Biomedicine Online* 23, 609–17.

Whittaker, A. (2011b) 'Cross-border assisted reproduction care in Asia: Implications for access, equity and regulations' *Reproductive Health Matters* 19, 37, 107–16.

Wilkinson, S. (2003) 'The exploitation argument against commercial surrogacy' *Bioethics* 17, 169–87.

Zanini, G. (2011) 'Abandoned by the state, betrayed by the Church: Italian experiences of cross-border reproductive care' *Reproductive Biomedicine Online* 23, 565–72.

8
Transplant Tourism

Thomas D. Schiano and Rosamond Rhodes

The key arguments of this chapter are as follows:

- Because of increasing numbers of patients requiring transplantation, more people are considering transplant tourism as a necessary option.
- Several studies show that complication rates are higher and graft and patient survival rates may be lower in transplant tourists returning to the US.
- Numerous professional transplant societies have produced consensus or position statements regarding transplant tourism, and these are summarised in the current chapter.
- Transplant tourism to the People's Republic of China has raised the issues and dilemma of the use of executed prisoners as donors.

Introduction

Organ transplantation saves lives and improves the quality of life for patients with end-stage organ failure. Advances in surgical techniques coupled with developments in immunosuppression have contributed to the growing success of organ transplantation and to ever-increasing graft and survival rates. Whereas the overall number of organs donated for transplantation has increased slightly year by year, the percentage of deceased donor organs that are donated has remained about the same. In spite of national and local efforts to increase public awareness and boost organ donation, the numerical increase in organ donation is not large enough to meet the growing demand. For example, in the US approximately 10 per cent of the patients on liver transplant waiting lists currently die each year because they do not receive a transplant organ (UNOS, 2011). It also means that some people with end-stage organ failure recognise that they are likely to die for lack of an organ or that they will have to wait many years for transplantation while living with pain and disability. The shortage of human organs for transplantation has created fierce competition between the transplant centres that are advocating for their patients: national and regional allocation policies have subsequently become highly controversial.

The problem of organ scarcity is compounded by the fact that more and more patients are now requiring transplantation. For example, greater numbers of patients are requiring liver transplantation due to hepatitis C (HCV), the epidemic of obesity and the resultant fatty liver disease, and the dramatic rise in hepatocellular carcinoma. Liver disease is now the ninth leading cause of death in the US and it is the most common cause of death in people with HIV. Similarly, increasing numbers of patients require kidney and heart transplantation and as time progresses more recipients of solid organ transplantation develop allograft failure and require retransplantation.

The shortage of organs for transplantation has repeatedly moved societies to reconsider their donation policies. In the US, when organ transplantation was a novel treatment, we started with a view that organ donation should be entirely optional – that no one should be encouraged to donate the organs of their deceased loved one and that no one should be required to make the request. As transplantation moved beyond the realm of the experimental, and became the standard of care, people noticed that a large number of organs that could have been used were never requested. This realisation moved legislators to abandon 'optional request' and move to a policy of 'required request'. Now, in the US whenever a suitable organ donor is being evaluated for brain death, the hospital is required to contact its local Organ Procurement Organization (OPO) so that a representative can make a request for the donation of transplantable organs from the family. Although this change in policy has increased the supply of organs, the demand still dramatically outweighs the supply. Several suggestions for further increasing the supply of organs have been put forward (Delmonico et al., 2002). One option would be to provide an incentive (for example, payment of funeral costs) to encourage families to donate the transplantable organs of deceased loved ones. Whereas this change could increase the rate of organ donation, people worry that it would both undermine altruistic donation and turn human organs into a commodity. Presumed consent is another policy option that has been adopted in some countries and is receiving serious consideration in others. If some version of presumed consent were accepted, upon the declaration of brain death, a patient's transplantable organs could be retrieved for transplantation unless the patient had previously documented a refusal of donation. In countries where such policies have been accepted, this change has led to a significant increase in organ donation. Yet, some resist such a move because they see it as coercive and undermining autonomous choice.

Organ allocation in the US

Currently, organ allocation in the US is administered by the Organ Procurement and Transplant Network (OPTN) in accordance with the 1984 National

Organ Transplant Act and with oversight for equitable allocation by the Department of Health and Human Services (DHHS). As of 28 December 2011 there were overall 112,740 people on solid organ transplant waiting lists. Different solid organs are matched to recipients using different criteria. For instance, kidneys are matched by human leukocyte antigen (HLA) allotype and livers are matched based on patient size and blood type.

In the US most organs are allocated to recipients at the local or regional level. This system of local priority has created geographical differences in the length of waiting time for patients and has spurred a national debate regarding the fairness of current organ allocation policies, many of which were first instituted in the 1980s. The consequent geographic disparities in waiting times for patients in the US are significant. For example, 15–17 per cent of patients listed for liver transplantation in New York die before they are allocated an organ, and kidney transplant patients can wait six to seven years for an organ, both much higher rates than in other regions of the US. The reason for this difference is likely to be multifactorial. It may be related to differences in the performance of local organ procurement organisations, comparative differences in the number of patients with end organ damage and the willingness for people to be donors.

Without a change in the current organ procurement policies that would significantly increase the overall supply of organs, the severe shortage of transplantable organs forces transplant programmes and procurement organisations to constantly make difficult decisions. Because of organ scarcity, many transplant programmes use extended donor organs, such as kidneys from older donors and livers from donors who are obese, have longer cold ischaemia times or have been exposed to viral hepatitis. Use of these organs that might otherwise have been rejected for use in transplantation may impact graft and patient survival rates and also increase the financial burdens on our health care system. These dire circumstances thus encourage some patients to become transplant tourists.

Equity in transplant organ allocation

Transplantation policies do not and should not treat all people equally: everyone is not given equal access to organs for transplantation (for example, one per person). Transplant organs, which are especially scarce and precious resources, are reserved only for those who need them. Principles for their just distribution aim at achieving equity (rather than equality) by treating relevant differences among those who need a transplant similarly. While there are many differences among candidates, the crucial policy problem becomes specification of the relevant differences and assignment of a relative priority to each of these relevant differences. When a policy gives certain irrelevant differences significant weight and when that assignment results in unequal treatment of similarly situated transplant candidates, the policy is unjust.

Most transplant programmes treat non-medical judgements about patients as irrelevant differences and, for the most part, resist the impulse to make blatant personal or relative judgements about recipient worthiness (Jonsen, 1998). This attitude reflects medicine's general commitments to the non-judgemental regard of every patient and a caring attitude towards each. These are professional commitments because they play an essential role in promoting the community's trust. We all want our doctors not to judge us harshly and to take good care of us, regardless of who we are and what we have done (Rhodes, 2007). For example, in wartime, doctors have a professional responsibility to treat all medically needy soldiers alike, those from their own army as well as enemy soldiers. Medicine's implicit attachment to these principles of non-judgemental regard and caring for all enables patients to bring themselves to doctors so that they can receive the benefits that medicine has to offer (Pellegrino and Thomasma, 1993). These professional positions on the appropriate physician attitude towards patients have translated into the transplant community's reluctance to judge recipient worthiness, recipient behavioural contribution to their present need, recipient age or even how much of a good life the patient has already enjoyed (Kamm, 2002).

By quantifying severity of disease, organ allocation policies can be seen as attempting to identify and to prioritise only medically relevant differences in need so that transplant candidates can be treated equitably. Transplant systems typically aim at establishing criteria for making uniform measurement for urgency of need (that is, how *soon* someone will die without the transplant, how *badly off* someone will be without it) so that patients who are listed for transplantation at different centres can be fairly assessed and compared (Kamm, 2002). Although arguments persist about how much weight should be assigned to each consideration, today's listing criteria are intended to reflect differences in urgency of need and they can be validated with clinical data and adjusted to reflect refined understanding of relevant medical factors that impact urgency. While the specific criteria and standards vary somewhat from organ to organ, because of immunological sensitivities and features specific to the survival of particular organs, these assessment instruments are supposed to quantify medical differences and, beyond these relatively objective criteria, leave priority to fairness as approximated by a rule of first-come-first-serve.

The MELD score

In the US in the late 1990s donor livers were allocated based on the severity of illness as documented by the patient's Child-Pugh Turcotte score (CPT), which quantifies the degree of encephalopathy, ascites, albumin, INR and bilirubin. Prior to relying on the CPT score, overall waiting time was a key factor in allocating livers. Basing organ allocation on waiting time had inherent problems. Some patients were listed years before they would need a transplant in order to give them an advantage at the time they actually needed a transplant. Thus, patients who were listed longer would be higher on the list as compared to patients who were severely ill but only recently listed.

In 1999 Institute of Medicine (IOM) published their report, *Organ Procurement and Transplantation: Assessing Current Policies and the Potential Impact of the DHHS Final Rule,* concluding that a more objective allocation system was necessary. The IOM recommended that the system should be based solely on laboratory testing to avoid the subjectivity of assessing degrees of ascites and encephalopathy, and that it should also minimise the emphasis on waiting time. These recommendations led to the adoption of the Model for End-Stage Liver Disease (MELD) scoring system in early 2002.

The MELD score has been shown to accurately predict the mortality of patients who are in urgent need of liver transplantation. The MELD score is based on measurements of three indicators of liver disease: (1) the patient's serum creatinine, which measures kidney function, (2) bilirubin levels, which measure the liver's excretion of bile and (3) INR, a measure of the liver's ability to make blood clotting factor. A formula using these factors tabulates a score ranging from 6–40; the higher the score the sicker the patient. Patients with high scores are higher on the national waiting list, and thus more likely to be allocated a transplant organ. Matching a liver recipient to a donor does not require tissue typing. As we mentioned above, for liver transplantation, the donor and recipient need only to have the same blood type and their organs have to be an appropriate size. This means that only the medically relevant factors of blood type, organ size and MELD score have to be considered in matching donors and recipients.

The MELD system has been useful in predicting survival without transplantation, thereby allowing a greater number of patients to obtain a graft. Reliance on the MELD score, however, creates another set of problems. Liver disease is complicated and urgency is difficult to discern. Some significant differences in patients' illnesses are not taken into account by the MELD score. The repeated changes in the approach to hepatocellular carcinoma (HCC) illustrate how the MELD system has been amended to more accurately reflect urgency of need. Typically patients with HCC have low MELD scores. Without being granted a variance they would not be able to obtain a transplant before the disease progressed to a point where transplantation was no longer feasible. Numerous studies have demonstrated that transplanted patients with HCC that meet specific tumour size criteria have excellent long-term disease-free survival. Patients with very small tumours were therefore initially given priority at the advent of the MELD era. Later it was decided that the priority granted to HCC patients disadvantaged other patients with more severe, immediately life-threatening chronic liver disease. Because the HCC patients could wait longer on the list for their transplant, the policies were amended so as to give priority only to patients with somewhat larger HCC.

Aside from urgency of need, listing criteria have focused on the medical judgement of likelihood of efficacy, that is, how likely it is that the patient will survive transplantation and that the organ and patient will remain viable for a significant period post-transplantation. Efficacy is a relevant consideration because deceased donor organs are scarce and justice requires that they be allocated so as to provide considerable benefit. Policies for organ distribution therefore evaluate patients for the likelihood of their long-term survival and the likelihood of post-transplant organ survival. When a potential recipient becomes so ill that the likelihood of survival is notably diminished, the patient is not listed for transplantation or is made inactive on the United Network for Organ Sharing (UNOS) list. Consideration of efficacy also requires that patients be evaluated for the likelihood of their adhering to rigorous post-transplant protocols so that the transplanted organ will not be lost to rejection. Typically, when a patient's history raises questions about the likelihood of adherence to a schedule of anti-rejection medications and post-transplantation medical monitoring, the patient is further examined and assessed by a psychiatrist or a social worker. Because adherence is an indication of efficacy, it is a reasonable and relevant medical consideration for the evaluation of individual patients.

Patients who need a transplanted organ overall have few options for improving their chances of receiving one. Live donation is one alternative, yet only about 10 per cent of recipients have a suitable donor, and for those who do, live donation involves attendant risks for the donor that have to be taken into account. Patients who have the wherewithal to seek multi-institutional listing may increase their chances of being transplanted by being listed at a centre in a region that has shorter waiting times. Although UNOS mandates that all patients be informed of the option of listing in additional regions, multi-institutional listing is often not feasible because of the travelling that is involved, health insurance restrictions or lack of financial resources.

Some facts about transplant tourism in the US

An analysis of the Scientific Registry of Transplant Recipients (SRTR) published in August 2007 identified 173 wait-list removals in the US, including 158 patients waiting for a kidney transplant, presumably because they had received an organ transplant at a foreign centre (Merion et al., 2008). The authors of this analysis estimate that an additional 200–335 patients had received transplants abroad for a total of 373–408 transplanted patients. About 75 per cent of these foreign transplants occurred between 2004 and

2006, and it appears that this number is continuing to increase. More than 40 per cent of these transplant tourists resided in New York and California. Although the majority of US transplant professionals frown upon transplant tourism, the practice violates neither current US law nor the National Organ Transplant Act (US Statute Large 1984, 98, p. 2339–48). In the US current UNOS policies even allow a small percentage of each centre's transplants to be allotted for foreign nationals (UNOS Policy 6.0, 2012). In essence, this policy does allow for transplant tourism within the US.

Transplant tourism in developing nations has been associated with significant problems for those who sell their organs. The money they are promised is paltry. Those who sell their organs are frequently swindled out of some of the procurement fee by middlemen and organ vendors. The surgery used to procure the organs and the post-transplant care may be substandard at some centres. Aftercare that should be provided to donors is typically not made available (Budiani-Saberi and Delmonico, 2008).

According to most studies, patients who receive organs as transplant tourists also experience significant medical problems. Patients who become transplant tourists typically learn about the option by word of mouth from friends or relatives. They have little reliable information about the quality of the programmes where they will receive a transplant. Typically, organ recipients do not receive adequate patient education either pre- or post-transplant, a crucial component of good medical care. There may also be poor communication throughout the process because of language barriers.

Like the paid donors involved in transplant tourism, the organ recipients, too, may be victims of sub-standard surgical techniques. In addition, their transplants may be compromised by poor organ matching, unhealthy donors and post-transplant infections. If they survive their transplants, they often receive inadequate post-surgical treatment. Patients may be discharged from the facility and encouraged to travel prematurely, and their immunosuppressive medication supply may be inadequate. When they return to their home transplant centre they often have inadequate records of what was done to them, little information about the organs they received or no records at all. All of these problems compromise their health, their lives and their transplanted organs (Ivanovski et al., 1997; Chugh and Jha, 2000; Sever et al., 2001; Higgins et al., 2003; Inston et al., 2005; Kennedy et al., 2005; Canales et al., 2006; Prasad et al., 2006; Bramstedt and Xu, 2007).

Several recent publications have highlighted the reality of poor outcomes in the returning transplant tourist. In one paper, Allam et al. (2010) reviewed the liver transplant outcomes of patients from several centres in Egypt and Saudi Arabia who had travelled to China to receive deceased donor transplants. Compared to patients being transplanted at their own centres, the investigators found decreased three year patient and graft survival rates, higher complication and hospitalisation rates, as well as higher rates of transmitted HBV infection. In another paper, Alghamdi et al. (2010) compared outcomes of 93 patients from their nephrology practice who obtained their kidney transplants outside of Saudi Arabia, 77 per cent in either Pakistan or the Philippines. The transplant tourist recipients had higher incidences of acute rejection, infectious complications and HCV transmission. Another study by Cha et al. (2011) compared their patients undergoing transplantation at Seoul National University Hospital to their patients who received a kidney allograft outside of Korea during the same time period. They found that amongst 87 of their transplant tourist patients there were much higher rates of acute rejection, infectious complications and hospitalisations as well as decreased graft survival.

Combined, the practices associated with transplant tourism translate into problems for the entire transplant community. Health professionals who work in transplantation worry that the transplant tourism industry could undermine our reliance on altruism for organ donation and undermine society's overall trust in transplant programmes. Ultimately, these unsavoury and unsafe practices could have a negative impact on future organ donation.

Transplant tourism and the People's Republic of China

Concern over the stigma associated with transplant tourism has extended to transplantation in the People's Republic of China (PRC). The Chinese government only recently admitted to using executed prisoners as the source of transplanted organs although it had been suspected for many years. Ethical questions have been raised about the judicial processes involved in these death sentences and about the voluntariness of the organ donation (Anonymous, 2006). In response to international consternation, the Chinese government has curtailed the practice of transplant tourism.

In 2007, the Chinese Ministry of Health published the Regulation on Human Organ Transplantation in an effort to regulate organ transplantation and create a legal and sustainable voluntary organ donation infrastructure (Huang, 2007). The number of yearly transplants in China has now surpassed the number performed in the US and the number is expected to continue climbing due to the population's high incidence of HBV and subsequent development of HCC. The majority of transplants are still performed using the organs of executed prisoners. This practice was not banned by the 2007 Regulation. It continues to be widely criticised internationally by human rights activists, ethicists and the transplant community. Although the Chinese Ministry of Justice has mandated that the removal of a prisoner's organs can proceed only if informed consent has been obtained from the prisoner or members of the prisoner's family, justifiable concerns about coercion and compliance with this ruling remain.

The number of transplants using organs from prisoners has actually decreased over the last several years in China. This change has exacerbated the need for live donor transplants. The Chinese government, however, discourages live donor transplantation and their universal health care insurance currently does not cover complications from live donor surgeries. Because of this and the fact that live organ donation is restricted only to relatives of the recipient, an illegal trade in organs from living donors has now developed in China (Huang et al., 2012).

Transplant tourists and transplant centres

The attitudes of transplant centres towards returning transplant tourists widely vary. Some centres provide care to patients whom they previously followed and others refuse to provide care for such patients. Biggins (2010) conducted an anonymous Internet-administered case-based questionnaire survey of transplant professionals in order to assess the influence of practices in the PRC on their patient care decisions. Of the respondents, 87 per cent considered procurement practices to be ethically sound in the US, but only 4 per cent considered the practices in the PRC to be ethically sound. The majority of respondents would provide post-transplant care for returning patients: 90 per cent would provide care for patients who returned from another domestic centre, 78 per cent would provide it for patients coming from another foreign country and 63 per cent would provide it for patients returning from a transplant in the PRC (Biggins, 2010).

The harsh reality of the organ shortage in the US means that all transplant professionals will be faced with the issue of returning transplant tourists. It also raises the question of whether patients should be informed about the option of transplant tourism when they are ineligible for a transplant or unlikely to be transplanted in their home country. The medical responsibility to put a patient's well-being before one's own, the commitment to beneficence and physicians' fiduciary responsibility all dictate that this information should be provided to patients who could benefit from it. Although coverage of post-transplantation medical care by insurance carriers has not been a major obstacle to returning US transplant tourists, patients should be instructed to confirm that they will be able to continue their benefits if receiving a transplant abroad (Boschert, 2007).

In dealing with sometimes desperate patients and families, the transplant professional must remain non-judgemental. Patients should face no disparagement because of their need to secure a transplant abroad. Although physicians may try to dissuade patients from becoming transplant tourists, patients should not be threatened with abandonment by a centre's refusal to provide care upon their return. Patients who do not meet reasonable criteria for transplantation in their home nation (for example, patients with an advanced malignancy) should be informed about the poor survival data associated with their diagnosis and dissuaded from seeking transplantation altogether.

Professional society statements and guidelines on transplant tourism

The growing concerns over transplant commercialism, organ trafficking and transplant tourism were the subjects of a 2008 summit convened by the Transplantation Society and the International Society of Nephrology that resulted in the Declaration of Istanbul (Steering Committee of the Istanbul Summit, 2008). The participants acknowledged that transplant tourism is a consequence of the global shortage of transplantable organs and recommended that all nations strive to ensure programmes are available to meet the transplant needs of their populations. They agreed that efforts to initiate or enhance deceased organ donation are essential in order to minimise the burden on live donors and that provision of care to the live donor is no less essential than taking care of the transplant recipient. The Declaration went on to emphasise that organ trafficking and transplant tourism should be prohibited because they violate the principles of equity, justice and respect for human dignity. The authors maintained that because transplant commercialism appeals most to the impoverished and other potentially vulnerable donors it contributes to injustice. Thus, they concluded that all types of advertising for donors as well as soliciting or brokering for organs should be banned.

Other societal statements

In April 2007 the International Society for Heart and Lung Transplant (ISHLT) issued a statement on accepting organs from executed prisoners. The ISHLT Statement on Transplant Ethics argued that the practice contravenes the principles of voluntary donation, provides perverse incentives to increase the number of executions and lays the judicial process open to corruption. The Society therefore concluded that members 'should discourage patients from seeking transplantation in countries where there is no external scrutiny' and assurance of ethical standards (ISHLT, 2007). The statement goes on to threaten sanction of those who fail to comply. It states that 'members who have been found to have contravened this ethical principle may have their rights and privileges as members suspended or removed'. In conformance with their stand, the ISHLT now requires a personal declaration that the authors adhere to these principles to accompany all submissions to their journal, the *Journal of Heart Lung Transplant* (ISHLT, 2007). The American Association for the Study of Liver Diseases (AASLD), the International Liver Transplant Society (ILTS) and the editors of *Liver Transplantation* have endorsed similar stances and created policies to uphold similar standards. They have taken positions against exploitation of donors and recovery of organs from executed prisoners and have condemned the use of paid living donors. They have also mandated that original publications should explicitly exclude the use of executed prisoners or paid donors, but they do not restrict basic research publications from transplant centres involved in such practices (Rakela and Fung, 2007). In the same vein, the UNOS Board of Directors issued a Statement on Transplant Tourism in June 2007 supported by the UNOS Ethics Committee (UNOS Board, 2007), and the Declaration of Istanbul on Organ Trafficking and Transplant Tourism condemned transplant tourism (Steering Committee of the Istanbul Summit, 2008).

The UNOS Statement does, however, acknowledge that transplant tourism exists because of a growing disparity between organ demand and supply. It therefore recommends that in emergency situations patients who have received a transplant abroad should be evaluated and treated according to the standard of care. Although it does not affirm an obligation for individual physicians to treat such patients in non-emergency situations (whether they were previously known to the transplant centre or not), the Statement does maintain that the medical community has the obligation to provide medical care for these patients.

While holding that medical care should not be withheld from those recipients who have chosen to receive transplants as 'tourists' from abroad, both the UNOS Statement and a similar declaration by the American Society of Transplantation (American Society of Transplantation, 2007) include an exemption from the duty to provide non-emergent care for those 'physicians who raise a conscientious objection'. Based on these official academically supported exemptions from professional responsibility, several transplant programmes have refused to render care to transplant tourists. Allowing physicians to refuse care to returning transplant tourists can make it awkward, unpleasant, difficult or impossible for these organ recipients to receive the ongoing expert medical care that they need.

In 2009 the American Society of Transplant Surgeons (ASTS) Ethics and Executive Committees endorsed the Declaration of Istanbul on Organ Trafficking and Transplant Tourism. The ASTS strongly agrees with the principles espoused in the Declaration: maximising the use of deceased donor organs, encouraging nations to share knowledge, protecting organ donors, ensuring equitable allocation of donor organs and preventing organ trafficking. The ASTS went further to suggest conducting trials of incentives for organ donation and to mandate that live donors be provided with insurance. Although a programme currently exists in the US which would provide a catastrophic insurance policy to live kidney donors, it has not been widely used. The ASTS proposed extending federal Medicare beneficiary eligibility to all live organ donors who lack private sector health care coverage (Reed et al., 2009).

Also in response to the Declaration of Istanbul the Canadian Society of Transplantation and Canadian Society of Nephrology in 2010 also issued a policy document. It listed several recommendations:

- All patients with end-stage organ failure should receive information about the pitfalls, increased complication rates and ethical concerns regarding transplant tourism and organ trafficking.
- Transplant professionals should not speculate on the specific transplant outcomes of specific institutions or countries since reliable standardised information is lacking.
- Patients should be apprised of the potential unethical treatment of individuals who sell their organs within unregulated systems in the developing world.
- Although transplant physicians have a fiduciary responsibility to act on behalf of their patients, their obligation does not include testing or prescribing medications in preparation for transplantation of a purchased organ.
- Physicians are obligated to care for transplant tourists in an emergent situation, but they may elect to defer care to another physician under non-emergent circumstances.

The First Latin American Bioethics and Transplant forum, sponsored by the Latin American and Caribbean Transplant Society, and all the transplant societies from subsidiary countries in 2010 issued the 'Document of Aguascalientes' which endorsed also the Declaration of Istanbul (Baquero and Alberu, 2011).

Should payment for donor organs be allowed

Organ donations are good because they alleviate suffering, save lives, conserve medical resources and save money. So long as the donation is voluntary, the expected benefits to the recipient are significant and the expected immediate and long-term burdens and risks to the donor are within a reasonable range, we consider living donor transplantation to be good, not morally objectionable. When emotionally related or good Samaritan living donors provide transplant organs to others who need them, the donations are altruistic and people find them praiseworthy.

When foreign nationals travel to developed nations like the US in order to receive an organ transplant, few people raise moral objections. The few objections that are voiced concern the allocation of organs to those who are outside of the region or the allocation of organs to those who are wealthy enough to travel and pay for their organ. These are matters of distributive justice that are relevant to the allocation of a community's scarce resource of deceased donor organs. These concerns do not, however, explain the objection to transplant tourism.

Yet, the problems of deception, exploitation and abuse of paid organ donors in developing nations, and the problems of risk to transplant tourists that we have described above, are significant. Clearly, no one condones the horrendous practices associated with transplant tourism that have been exposed in developing nations. That said, the remaining theoretical question is whether payment for organs could be ethically allowed under different circumstances. In other words, if the transaction were voluntary involving no force or deception, the expected benefits to the recipient were significant and the expected immediate and long-term burdens and risks to the donor were within a reasonable range, would transplant tourism be an acceptable practice? To assess the morality of organ sales under these circumstances, imagine an environment in which paid organ donors received a very significant sum for their donation (for example, twice the average annual salary in their home country), and the terms of their contracts were scrupulously upheld and enforced by their government. Further imagine that the medical conditions of their organ donation received careful attention and oversight, including thorough donor evaluation to rule out any donors for whom donation might be an unreasonable risk, responsible medical care and access to needed future health care.

In some circumstances there are powerful reasons for prohibiting payments. For example, medical credentials should not be sold because society needs to rely upon health professionals for their clinical expertise. Judges should not be allowed to sell their verdicts because that would pervert the legal justice system and undermine social stability. In other circumstances there are no powerful reasons for prohibiting payments. If there are no persuasive reasons for opposing payments, they should be allowed. The traditional reasons for restriction on liberty are that the prohibition would avoid badness, prevent harm to others or prevent wrongs to others.

For the past 25 years or so, people have offered objections to payment for organs by presenting reasons of each of these sorts. Rather than discussing the articles by individual critics of the practice, in the following section we shall examine the kinds of objections in turn, in order to determine whether any of them amount to persuasive reasons for prohibiting payment for organs.

Badness as the reason to prohibit payment for organs

Some people see monetary payment as something dirty. For them the problem is simply that 'filthy lucre' is involved. Yet, monetary exchanges offer many advantages. Money facilitates voluntary transactions and allows people to exchange the things that they have for what they need. This enables people to get things that they want. In this way, voluntary exchanges of money promote autonomy by enabling people to make choices that express their values and priorities and support liberty by allowing people the freedom to do what they like with what they have. Money also augments well-being by allowing people to determine for themselves what will make them better off.

Some say that payment for organs should not be allowed because selling body parts would be alienation of a part of oneself. They seem to overlook the fact that their objection would apply equally to unpaid organ donors and to amputation. Also, as the case of Aron Ralston, the subject of the film '127 Hours' who cut off his own arm, demonstrates, there are circumstances in which the goods achieved and the harms averted can justify separating a part of oneself.

Some object to payment for organs because to them selling body parts is commodification, treating a person as a thing or as property that can be bought or sold. They invoke Immanuel Kant's claims that the dignity of persons should be respected and that people should not be treated as a means only. Yet, for Kant, the seller is a person who should take responsibility for her/his own actions. That would not rule out a decision to donate an organ with or without payment. As Kant might reply, others should treat both organ donors and organ sellers with respect for their autonomy, their judgement, their choice and their courage.

Some contend that payment for organs should not be allowed because allowing markets would change the nature of donation. It is certainly true that if payment were to be allowed, what has been priceless would then have a price. There was a time when mothers nursed their infants, and families directly provided education and caregiving for their offspring and family members who needed care. Today, mothers feeding their infants formula or families sending their children to school or hiring caregivers for family members who need constant care are not regarded as morally problematic. Changes can be better in some respects and worse in others, but the fact that they are changes does not make them wrong. No one has the right to deny others freedom simply because they prefer to maintain the *status quo*.

Harm to others as the reason to prohibit payment for organs

Some have argued that payment for organs should not be permitted because it would increase pressure on people to earn money by selling their organs. Pressure in and of itself, however, is not a moral problem. People's own values and priorities press them to make certain choices. Also, donor families may exert tremendous pressure. Whether pressure to sell is greater or weaker than the pressure to give is an empirical question, but in neither case it does not explain why selling organs should be prohibited.

Some maintain that payment for organs should not be allowed because introducing paid organ donors would decrease altruistic donation and, thus, not provide a net benefit in organ availability. This again is an empirical question. It is easy to imagine that some wealthy people who need an organ would prefer paying for one over asking a family member to donate and some potential donors will feel relieved of a responsibility to donate if their family member had access to an organ by paying for it. If the objection here is that recipients would be harmed by a net loss in transplant organs, it should be noted that it is easier to alter a reimbursement scale to match recipient needs than it is to increase altruism.

Wrongs to others as the reason to prohibit payment for organs

People can be wronged without being made worse off or harmed in some way. Several arguments against payment for organs are based on a claim that people will be wronged by the practice.

One argument against payment for organs is that poor people would sell an organ. Opposition on this ground encompasses three objections that call for separate consideration. Would a disproportionate number of poor people accepting payment for transplant organs be unjust? It certainly can be expected that poor people would be the ones most likely to sell an organ. But unequal participation does not make an activity unjust. Prohibiting payment would deny poor people options that would increase their well-being. Furthermore, prohibiting payment is paternalistic. Without adequate justification, it expresses disrespect. There are many activities that poor people will choose to do to earn money that wealthier others would eschew. Unless the poor were paid too little for their organ donation, it is hard to see how allowing them to earn money that way would be unjust.

Others argue that payment for transplant organs would wrong people because it involved an unjust threat of harm that made the offer coercive. The offer of a benefit, such as a payment, is not coercion but inducement. If inducements undermine autonomy, then no payments to poor people (or others) can be ethically acceptable. If inducements do not undermine autonomy in other circumstances, then there is no obvious reason to suppose that payments for organs would be different.

Still others argue that a disproportionate number of poor people accepting payment for transplant organs is exploitation. Exploitation involves one person placing another in great danger and thereby receiving a disproportionately large personal benefit. Yet, managed according to the standards of current developed-nation living donor transplantation, the risks to a kidney seller would be reasonable and paid organ donors would not be subjected to greater danger than other living donors. Furthermore, paying a significant sum would avoid even the appearance of exploitation.

Conclusion

Because end-stage organ failure threatens death, and because of the ongoing organ shortage, increasing numbers of patients may resort to transplant tourism. Transplant teams must be open to this idea and provide ongoing education and support for them. People with chronic organ failure and end-stage disease already are suffering from a loss of self-esteem and independence, physical symptoms, financial pressures, anxiety and depression. To consider leaving the transplant centre that they have come to trust, to travel thousands of miles away from their homes and families, and to deplete their savings, and to subject themselves to major surgery by a team in an unfamiliar country where people speak another language are indeed stressful. Although none of us condone all of the current practices associated with transplant tourism, patients who try to save their lives by resorting to transplant tourism need information and the compassionate care that other transplant patients need.

People who see that their best option for securing their well-being and the good of their loved ones is in selling an organ need our protection. Our responsibility is to institute the social structures that will safeguard them from the deception, exploitation and abuse that paid organ donors are subjected to today. In the US and other developed nations, paid gestational surrogates are free to earn money from their valued service and they are protected from abuse because their medical care is carefully monitored and their contracts are enforced. Societies where transplant tourism thrives today should institute the measures that assure that those who are to receive payment for donating organs are protected. As a legal practice that is regulated and supervised with attentive oversight, potential donors can be carefully screened and provided with optimal care. These measures would assure both donor freedom and donor safety.

References

Alghamdi, S. A., Nabi, Z. G., Alkhafaji, D. M., Askandrani, S. A., Abdelsalam, M. S., Shukri, M. M., Eldali, A. M., Adra, C. N., Alkurbi, L. A. and Albaqumi, M. N. (2010) 'Transplant tourism outcome: A single center experience' *Transplantation,* 90, 184–8.

Allam, N., Al Saghier, M., El Sheikh, Y., Al Sofayan, M., Khalaf, H., Al Sebayel, M., Helmy, A., Kamel, Y., Aljedai, A., Abdel-Dayem, H., Kenetman, N. M., Al Saghier, A., Al Hamoudi, W. and Abdo, A. A. (2010) 'Clinical outcomes for Saudi and Egyptian patients receiving decreased donor liver transplantation in China' *American Journal of Transplant,* 10, 1834–41.

American Society of Transplantation (2 March 2007) 'Position statement on transplant tourism' http://www.a-s-t.org/sites/default/files/legacy_pdfs/public_policy/Declaration_of_Istanbul.pdf date accessed 4 May 2012.

Anonymous (2006) 'Organ sales "thriving in China" ' *BBC News* 27 September 2006, http://news.bbc.co.uk/2/hi/asia-pacific/5386720.stm date accessed 4 May 2012.

Baquero, A. and Alberú, J. (2011) 'Documento de Aguascalientes. Ethical challenges in transplant practice in Latin America: The Aguascalientes document' *Nefrologia,* 31, 275–85.

Biggins, S. W. (2010) 'Supply and demand in transplant tourism: Disclosure duties of the transplant physician and our global transplant community' *Liver Transplantation,* 16, 2, 246–7.

Boschert, S. (2007) ' "Transplant Tourists" pose ethical dilemmas for U.S. physicians', *GI and Hepatology News,* 11.

Bramstedt, K. A. and Xu, Jun (2007) 'Checklist: Passport, Plane Ticket, Organ Transplant' *American Journal of Transplantation,* 7, 1698–1701.

Budiani-Saberi, D. A. and Delmonico, F. L. (2008) 'Organ trafficking and transplant tourism: A commentary on the global realities' *American Journal of Transplant,* 8, 925–9.

Canales, M. T., Kasiske, B. L. and Rosenberg, M. E. (2006) 'Transplant tourism: Outcomes of United States residents who undergo kidney transplantation overseas' *Transplantation,* 82, 1658–61.

Cha, R. H., Kim, Y. C., Oh, Y. J., Lee, J. H., Seong, E. Y., Kim, D. K., Kim, S. and Kim, Y. S (2011) 'Long-term outcomes of kidney allografts obtained by transplant tourism: Observations from a single center in Korea' *Nephrology,* 16, 672–9.

Chugh, K. S. and Jha, V. (2000) 'Problems and outcomes of living unrelated donor transplants in the developing countries' *Kidney International,* 57, S131.

Delmonico, F. L., Arnold, R., Scheper-Hughes, N., Siminoff, L. A., Kahn, J. and Youngner, S. J. (2002) 'Sounding Board: ethical incentives – not payment – for organ donation' *New England Journal of Medicine,* 346, 2002–5.

Higgins, R., West, N., Fletcher, S., Stein, A., Lam, F. and Kashi, H. (2003) 'Kidney transplantation in patients traveling from the UK to India or Pakistan Letter' *Nephrology Dialysis Transplantation,* 18, 851–2.

Huang, J. (2007) 'Ethical and legislative perspectives on liver transplantation in the people's Republic of China' *Liver Transplantation,* 13,193–6.

Huang, J., Millis, J. M., Mao, Y., Millis, M. A., Sang, X. and Zhong, S. (2012) 'A pilot programme of organ donation after cardiac death in China' *Lancet,* 379, 862–5.

International Society for Heart and Lung Transplant, Statement on Transplant Ethics, Approved April 2007, https://ww.dafoh.org/ISHLT_-_Statement_on_tran.php date accessed 4 May 2012.

Institute of Medicine (IOM) (1999) *Organ Procurement and Transplantation: Assessing Current Policies and the Potential Impact of the DHHS Final Rule* (Washington, D.C: National Academy Press).

Inston, N. G., Gill, D., Al-Hakim, A. and Ready, A. R. (2005) 'Living paid organ transplantation results in unacceptably high recipient morbidity and mortality' *Transplant Proc,* 37, 560–2.

Ivanovski, N., Stojkovski, L., Cakalaroski, K., Masin, G., Djikova, S. and Polenakovic, M. (1997) 'Renal transplantation from paid, unrelated donors in India – it is not only unethical it is also medically unsafe' Letter, *Nephrology Dialysis Transplantation,* 12, 2028–9.

Jonsen, A. R. (1998) *The Birth of Bioethics* (New York: Oxford University Press).

Kamm, F. M. (2002) 'Whether to discontinue nonfutile use of a scarce resource' in R. Rhodes, M. Battin, and A. Silvers (eds) *Health Care and Social Justice* (New York: Oxford University Press).

Kennedy, S. E., Shen, Y., Charlesworth, J. A., Mackie, J. D., Mahony, J. D., Kelly, J. J. and Pussell, B. A. (2005) 'Outcome of overseas commercial kidney transplantation: An Australian perspective' *Medical Journal of Australia,* 182, 224–7.

Merion, R. M., Barnes, A. D., Lin, M., Ashby, V. B., McBride, V., Ortiz-Rios, E., Welch, J. C., Levine, G. N., Port, F. K. and Burdick, J. (2008) 'Transplants in foreign countries among patients removed from the US transplant waiting list', [2007 SRTR report on the state of transplantation] *American Journal of Transplant,* 8, 988–96.

Pellegrino, E. D. and Thomasma, D. C. (1993) *Virtues in Medical Practice (*New York: Oxford University Press).

Prasad, G. V., Shukla, A., Huang, M., D'A Honey, R. J. and Zaltzman, J. S. (2006) 'Outcomes of commercial renal transplantation: A Canadian experience', *Transplantation,* 82, 1130–5.

Rakela, J. and Fung, J. J. (2007) 'Liver transplantation in China' *Liver Transplantation*, 13, 182.

Reed, A. I., Merion, R. M., Roberts, J. P., Klintmalm, G. B., Abecassis, M. M., Olthoff, K. M. and Langnas, A. N. (2009) 'The Declaration of Istanbul: Review and commentary by the American Society of Transplant Surgeons Ethics Committee and Executive Committee' *American Journal of Transplant*, 9, 2466–9.

Rhodes, R. (2007) 'The Professional Responsibilities of Medicine' in R. Rhodes, L. Francis and A. Silvers (eds) *The Blackwell Guide to Medical Ethics* (USA: Blackwell).

Sever, M. S., Kazancioğlu, R., Yildiz, A., Türkmen, A., Ecder, T., Kayacan, S. M., Celik, V., Sahin, S., Aydin, A. E., Eldegez, U. and Ark, E. (2001) 'Outcome of living unrelated (commercial) renal transplantation' *Kidney International*, 60, 1477–83.

Steering Committee of the Istanbul Summit, *The Declaration of Istanbul on Organ Trafficking and Transplant Tourism*, meeting convened by the Transplantation Society and International Society of Nephrology, Istanbul, Turkey, 30 April through 2 May 2008.

Transplantation Society (2007) 'Policy on interaction with China' *Transplantation*, 84, 292–4.

United Network for Organ Sharing Policy 6.0. (1 Sepetember, 2012) Transplantation of non-resident Aliens, http://optn.transplant.hrsa.gov/PoliciesandBylaws2/policies/pdfs/policy_18.pdf date accessed 4 May 2012.

UNOS Board of Directors Resolution Regarding Transplant Tourism (26 June 2007) http://www.unos.org/about/index.php?topic=newsroom&article_id=2194 date accessed 4 May 2012.

US Statute Large (1984) 98, 2339–48.

UNOS Data (2011) *United Network for Organ Sharing*, http://www.unos.org date accessed 4 May 2012.

9

The European Cross-Border Patient as Both Citizen and Consumer: Public Health and Health System Implications

Tomas Mainil, Matt Commers and Kai Michelsen

The aims of this chapter are as follows:

- To provide a deeper explanation for why the EU patient in cross-border care is not only a consumer but also a citizen, by a more thorough explanation of how and why the EU context is unique.
- To create a typology of cross-border care that builds on existing ideas but incorporates the dimensions raised above. The analysis of the regulatory history of cross-border care in Europe also raises a number of important policy and political questions.
- To explore a number of scenarios for the future evolution of cross-border care in the European context.
- To recommend that a future research agenda monitors the application by the EU members of the Patients' Rights Directive (2011/24/EU), as well as the monitoring of transnational health care development.

Introduction

In total, the quantity of 'medical travel', 'medical tourism' or cross-border 'patient mobility' and service provision is still limited, for example, in relation to total health expenditures and service provision. Currently, issues of cross-border social security, cross-border mobility of patients and services are of special relevance for tourist regions, regions attracting retired persons and border regions. However, the question of how economic and political dynamics will shape these kinds of cross-border affairs remains open.

Our chapter in many ways takes its starting point from a recent article by Carrera and Lunt (2010) which provides a European perspective on 'medical tourism' (Snyder et al., 2011), 'medical travel' (Ormond, 2011; Turner, 2011) and 'patient mobility'. Carrera and Lunt have suggested that the EU patient in cross-border care can be characterised as not only a consumer – as with cross-border care patients around the world – but also to a significant extent as a citizen. We agree with this characterisation.

While we see great value in Carrera and Lunt's article, we believe in general that their focus on the specific arena of 'medical tourism' implies the need for a broader perspective. We believe that the discussion and analysis of the patient as both citizen and consumer in the EU needs to be related to the patient perspective rather than to the industry perspective of medical tourism. This is because most of the regulations that make the EU patient unique in this regard (that is, the reasons that make the EU patient as much a citizen as a consumer) have to do with those who become patients while abroad (for example, those permanently residing in countries other than the country of affiliation, those living in border regions) and not with medical tourism (that is, tourism for the sake of medical consumption) as such.

In this chapter, we attempt to provide a deeper explanation for why the EU patient in cross-border care is not only a consumer but also a citizen. By providing a more thorough explanation of how and why the EU context is unique, we hope to contribute to the theoretical foundation for discussion and debate of policy alternatives for structuring cross-border care internationally and in the EU in the coming years. This means that we want to establish a link between 'medical travel', 'medical tourism', 'patient mobility' and its consequences for EU populations and the provision of health services. As a limitation, we primarily focus on the ways that health care is consumed by EU citizens within the EU and the legal context that surrounds that consumption. We do not extensively address issues relating to the mobility of European patients who seek care outside of the EU or the mobility of non-European patients who obtain care within the EU.

The chapter starts with references to the regulations framing medical travel, medical tourism and patient mobility within the EU, to illustrate the dual character of patients as citizens and consumers. On the one hand, cross-border affairs are regulated by mechanisms of social coordination. On the other hand, they are regulated by various aspects of European law, most notably the Directive on the Application of Patients' Rights in Cross-Border Healthcare, also known as the 'Patients' Rights Directive' (2011/24/EU). These two sides should complement each other. As these two regulatory lines reveal, there are many ways in which European citizens are not only consumers but citizens when they obtain care in countries other than their own. This analysis suggests a few meaningful ways in which existing typologies of cross-border care can be further refined. In the following section, we attempt to provide a typology of cross-border care that builds on existing ideas but incorporates the dimensions raised above. The analysis of the regulatory history of cross-border care in Europe also raises a number of important policy and political questions. In the section 'Towards transnational patients and services?', we explore a number of scenarios for the future evolution of cross-border care in the European context.

European cross-border patient as citizen and consumer: The regulatory framework for cross-border health care consumption and provision in the EU

Arguably, the legal context in which cross-border health care in Europe takes place has primarily been given form over time more by economic interests, legislation and jurisprudence than health policies at EU or Member State level. In some sense, therefore, it is a paradox that cross-border care within the EU is defined as much by citizens' rights as consumer forces.

In many ways, EU citizens consuming health care in a Member State other than their own are no different than any patient travelling for care throughout the world. The European patient may indeed travel to another Member State as a consumer – as many do – to obtain services that are cheaper, better or more accessible in a Member State other than their own. Laugesen and Vargas-Bustamente (2010) have called EU citizens' consumption of cheaper dental care abroad as an example of 'complementary exit', and specific cases (for example, such as the landmark ECJ cases Peerbooms and Watts) in which Europeans have sought better or more accessible care abroad as examples of 'duplicative exit'.

This process is arguably highly consistent with the original economic intentions and treaty agreements at the heart of the EU and the process of integration that the Union embodies. The EU was founded to secure the free flow across borders of people, money, goods and services (Official Journal of the European Union, 1992 C 191 1, 31 ILM 253). Article 119 of the TFEU guarantees EU citizens 'an economic policy... conducted in accordance with the principle of an open market economy with free competition' (Official Journal of the European Union, 2008 C 115/47).

It seems to be not particularly surprising, therefore, that the legal context for cross-border care in Europe was primarily given form by decisions – political but also jurisprudential – taken for economic reasons. Yet it might at a first look be surprising that, given that economic forces have given the EU context of cross-border care the majority of its form, rights based on citizenship – rather than ability to pay as a consumer – define the lion's share of the possibilities and limitations of that context. There are two major regulatory threads that have given rise to this uniqueness in the European context of cross-border health care. In this section, we provide a brief history of the evolution of policies and laws that provide a unique landscape for the consumption of cross-border health care in Europe.

The mechanisms of social coordination

As a consequence of European integration, policy makers were confronted with the dilemmas created by the fact that two major groups of European citizens were seeking health care in countries other than their own. The first group was workers – at all levels, from professional to working class – who were employed in a Member State other than their own, as well as their families who at times relocated to accommodate that employment abroad. The second group was comprised of pensioners who relocated after retirement for part or all of the year from their Member State of origin to another EU Member State.

In 1957, the Treaty of Rome (1957) (EEC 25 March 1957, 298 UNTS 3) established the European Economic Community (EEC). Increased movement of persons from one state within the EEC to another soon became the target of European regulation. Regulations 1408/71 and 574/72 as well as revisions 883/2004 and 987/09 secured the rights of workers, their family members as well as pensioners to contribute to and draw from social security systems in Member States other than their own. Also, these regulations established and over time have clarified and strengthened the ability of EU citizens to receive reimbursement for planned health care in Member States other than the one in which they pay social security contributions. They are treated in line with the statutory system of the EU Member State in which the services are delivered, at the expense of the EU Member State in which they are insured (Palm and Glinos, 2010, p. 515). But the situations in which reimbursement was regulated have been specified, and for example, for elective hospital services patients have had to ask their health insurances in advance for permission, while the insurer had the right to withhold the permission.

Here the situation of EU Member States differs significantly, for example, from the US. US citizens do not accrue US Social Security benefits when living and working abroad, nor can foreign earnings ever be applied back to an individual's accrued balance within US Social Security. Medicare, with extremely limited exceptions, does not reimburse care provided outside of the US (CMS, 2010). Hence, the US consumer of cross-border care does so most often exclusively as a consumer – one who is privately insured or pays out of pocket. He enjoys no 'right' to this care as a citizen of the US or Medicare insuree. This contrasts with European cross-border health care consumers, some of whom could – even from the early 1970s – base their claim to reimbursement in their rights as a citizen of a European Member State in a professional field of social security schemes.

Path to the directive on the application of patients' rights in cross-border health care

The process of establishing and clarifying the rights of migrant workers and their families and pensioners to cross-border health care occurred relatively early after – and on the basis of the competencies established at EU level by – the Treaty of Rome. However, a second line of establishing the right of EU citizens to obtain health care in Member States other than their own developed. This line of regulation ultimately led to the quite recent passage in 2011 of the Directive on the Application of Patients' Rights in Cross-Border Healthcare (2011/24/EU).

One of the early and most important milestones in the history of this development took place in the mid-1990s. At that time, at least two citizens of Luxembourg, Kohll and Decker, made health care expenditures outside of Luxembourg and declared those costs to their domestic insurer. When the claims were rejected because they were related to expenditures outside of Luxembourg, each gentleman sued the insurer. Each claimed that under European treaty law, he should be entitled to full reimbursement for expenditures made anywhere within the EU. Ultimately the European Court of Justice, which heard the cases together, decided in favour of the plaintiffs (Judgements Kohll and Decker, 28 April 1998, cases C-120/95 and C-158/96 (1998).

These and other cases soon led the European Commission to propose a Directive on Health Services within the Internal Market (Commission of the European Communities, 2008). The Commission's stated purpose in proposing this Directive was to clarify many aspects of cross-border health care in Europe – not only the patient's rights, but rights relating to providers, insurers and others as well. However, the proposal of a directive with such a broad scope proved impossible to implement. Instead of new law, a new round of reflection was the result. Robert Madelin chaired a high-level group starting in 2004 to discuss a wide variety of issues related to cross-border health care in Europe. This in turn fed into an open consultation of the Member States that took place in 2006.

The Patients' Rights Directive 2011/24/EU, which took force on 4 April 2011, provides a number of new and important assurances to citizens seeking cross-border care within the EU. Generally, under the Directive citizens of the EU are now guaranteed that care they consume in a Member State other than their own will be reimbursed at the rate that it would have been reimbursed in their own Member State (that is, the Member State of affiliation; this is an important difference to mechanism of social coordination, see above). This reimbursement may be paid directly by the Member State of affiliation to the care provider. Member States do retain the right to preauthorise care consumed outside the Member State of affiliation, though these circumstances are quite limited – much more limited than under the mechanisms of social security coordination. The Directive is expected to complement the rights established in Directive 993/2004.

The Patients' Rights Directive 2011/24/EU entails some regulations which have to be implemented by the Member States within the next years, addressing, for example, national contact points, information by health care providers, transparent complaints procedures and mechanisms, systems of professional liability assurance, data protection, pricing issues and reimbursements (Art. 4). The Member States are asked for mutual assistance and cooperation in setting standards and developing guidelines on quality and safety and the exchange of information. They shall facilitate cooperation in cross-border health care provision at regional and local level as well as through ICT and other forms of cross-border cooperation, and deliver information about the right to practice (Art. 10). Further issues address the recognition of prescriptions issued in another Member State (Art. 11), the establishment of European reference networks (Art. 12), health services for rare diseases (Art. 13), voluntary networks connecting national authorities responsible for eHealth (Art. 14) and Health Technology Assessments (Art. 15).

It has also been decided that the Commission will produce reports on patient flows, financial dimensions of patient mobility, implementation of Article 7 (9) and Article 8 and on the functioning of the European reference networks and national contact points. The Member States shall provide assistance and information (Art. 20).

With passage of the Patients' Rights Directive 2011/24/EU, citizens of EU Member States have become further distinguished from citizens of countries outside the EU when it comes to the consumption of cross-border health care. The EU is now at least 40 years into a process of continual expansion of citizens' rights when it comes to the consumption of such care. It is now roughly accurate to assert that the rights of EU citizens to obtain care in EU Member States other than their own are quite similar to the rights applying for citizens within their own country's health care system. But at the same time, it is still an open question if or what kinds of dynamics in cross-border mobility will follow the implementation of the regulation. The required reporting activities will hopefully deliver more systematic information and knowledge about cross-border mobility and respective developments in the coming years.

Types of patient mobility in the EU

The described EU regulations have been taken into account in the following development of a typology for medical travel, medical tourism and patient mobility. The development of this constructed typology (based on several authors) originates in the fact that Carrera and Lunt's consumer/citizen typology does not capture the full complexity of patient mobility, and that in our opinion, the terminology (medical travel, medical tourism and patient mobility) itself needs further clarification. Building on this line of reasoning, this would also include an assessment of policies to implement the right services to support patients and services providers in cross-border issues.

It is perhaps consistent with the complexity of the patient mobility phenomenon that different scholars have suggested several typologies to describe and categorise it. Carrera and Lunt, and their concept of combined citizenship and consumerism incorporated by the EU patient, is a unique contribution but does add to the work of others. One of the more complete typologies is that of Glinos et al. (2010). This typology was originally conceived from a demand driven impetus: The motivations of patients (familiarity, perceived quality, affordability and availability) were combined with the question of whether services would be covered by social security or health insurance systems or not. However, the focus on patient mobility limits the radius of this typology. Instead, we like to think in terms of mobile settings, where there is an integration of the situation in which the patient resides and how the professional constellation reacts to this situation. Evidence seems to support the idea that cross-border health care can function if every stakeholder is integrated, but that there is a gap between the reality and the ideal situation, which raises challenges for the future.

A second contribution to the proposed integrated framework we would like to suggest is the exit patient mobility typology of Laugesen and Bustamante (2010) which also seeks to argue in global terms rather than on an EU or US level. They derived types of international patients and the reasons why they want to exit their home country (sending context): They formulate it as follows:

> Four types of patient mobility are defined: primary, complementary, duplicative, and institutionalised. Primary exit occurs when people without comprehensive insurance travel because they cannot afford to pay for health insurance or directly finance care, as in the United States and Mexico. Second, people will exit to buy complementary services not covered, or partially covered by domestic health insurance, in both the United States and Europe. Third, in Europe, patient mobility for duplicative services provides faster or better quality treatment. Finally, governments and insurers can encourage institutionalised exit through expanded delivery options and financing. Institutionalised exit is developing in Europe, but uncoordinated and geographically limited in the United States.

Although this typology also takes a global perspective on patient mobility, it does not start from a mobile setting; rather, it argues for patient perspectives.

Mainil et al. (in press), taking a global perspective, aligning cross-border health care and medical tourism following a logic of access to health care, link the patient motivations of availability and familiarity with their own criteria to typify international patients. As such, the geographical proximity/distance to be covered by the patient is related to how available a medical service abroad is, by taking (relative) proximities and distances into account. Cultural proximity/distance of the foreign health system can be linked with how familiar that system is for the patient. Finally, the searching efforts that are deployed by the patients are based on a mix of both familiarity and availability: high levels of availability and familiarity decrease searching effort. These criteria offer the possibility of reasoning in terms of a duality of patient types.[1] The two types represent (1) high levels of the criteria and (2) low levels of the criteria.

In order to create an integrated typology, taking the different settings and dimensions of proximity and distance into account, one could combine the mindsets of Glinos et al., Laugesen and Bustamente, Carrera and Lunt, and Mainil et al. The latter show that if EU patients cross national borders, this phenomenon is often referred to as cross-border patient mobility, described by Glinos et al. as 'at a minimum involves a patient who travels to another country for the purpose of receiving planned health care'. A clear focus on the demand side of health care occurs here. In other parts of the world, such as the Asian and American continent, medical tourism is the term which is mostly used, as defined by Snyder et al., to describe 'a growing industry that involves patients intentionally travelling abroad for non-emergency medical services'. Here a supply side logic seems to be the semantic focus. Medical travel as a term is preferred by some scholars (Ormond, 2011; Turner, 2011) above medical tourism, probably because it excludes the nature of tourism, as being a purely supply-driven, leisure-based, commercialised sector, far away from the basics of pain, healing and providing care to patients, characteristics present in the health care sector. Patient mobility incorporates types of patients which are not classical international patients, sketched by the medical travel industry. An example can be found in the retired senior citizens from the UK, residing in Spain and in need of health care services (Legido-Quigley et al., 2012).

The term transnational health care (THC) could be used to better reflect patient mobility at a time when coordinated actions and enhanced structure and formality take place in cross-border care and medical tourism services. THC can be recognised, therefore, by the existence of extended global and local professional as well as provider – third-party payer– networks within the provision of health care services. Within established THC structures, patients would have to possess the possibility to make an appropriate and knowledgeable choice on the basis of the provided services within these networks to go abroad and to receive health care.

In this framework, we distinguish two types of patient settings: cross-border access searcher settings (CBASs) and trans-border access seeker settings (TBASs). CBASs are a prototype of patients who do not have to cope with large distances or cultural shifts. CBASs search for access in a regulated health care system in a proximal country (for example, the Dutch using the Belgian health care system). Or if no reimbursement is possible, they estimate if it is still worthwhile to pursue their treatment. CBASs have the possibility to be embedded in a safe social or private security system. This means that CBASs are centred around three criteria which makes them specific as a type:

1. The proximity and the limited travel distance inherent to this proximity (hence cross-border).
2. The cultural proximity (Bell et al., 2011): the medical culture they are encountering in the other health system is no different from their own, with medical staff who are recognisable.
3. The challenge of searching for health care in another proximal country is perceived to be of a reasonable scale: in the case that insurance schemes are a provider, they can offer information to their clients. If this is not the case it is still reasonable to search for information in a proximal country (hence access searchers). On the basis of these criteria, we estimate that availability (Glinos et al., 2010) and familiarity are perceived as high.

In contrast, TBASs are patients who have to travel more, more regularly pay out of pocket and encounter cultural differences within the context of the host (hospital) and guest (patient). TBASs are motivated to find alternative health care options, including how they can be treated abroad (for example, patients from Arab states seeking health care in Germany – Private), but also covered specialised treatments for EU patients are part of this type (for example, an Italian patient to be treated in Belgium – Public). This means that TBAS is also characterised by criteria:

1. The larger distance to travel makes it harder and more risky to undertake such health-related travel (hence trans-border).
2. There is also cultural distance (as opposed to cultural proximity (Bell et al., 2011)) between the mindset of the patient and the receiving (medical) context.
3. It involves for the patient a seeking exercise in stipulating which receiving context or health system they are going to choose (hence access-seeking). Taking into account these criteria, we estimate that availability and familiarity are experienced as lower, than in the case of CBASs.

The duality between TBASs and CBASs, on the basis of the three criteria of geographical distance, cultural distance and searching effort, is based on their different needs for supportive services; as such, the criteria position CBASs and TBASs at either end of a continuum: one end representing high ranges of the criteria and the other end low ranges of the criteria, respectively, with a full range between them. The supportive need for services is much higher in the case of TBASs than it is in the case of CBASs. The fact that these two types are at both ends of this continuum obviously means that other combinations of the criteria are present, but for didactical reasons they are not elaborated upon. Supportive services could be defined as cultural, informational, financial and logistic services, as opposed to medical services.

The added value of this typology is that it tries to connect on a meta-level several author-based conceptual exercises. The typology combines them in order to grasp the complexity inherent to patient mobility. In typifying two types of international patients (CBAS and TBAS), one could link them to the concept of public and private health care delivery. At the same time, public health care delivery relates to the practice of cross-border health care (CBHC) and citizenship, whereas private health care delivery refers more to medical tourism and consumerism. Finally, this public/private divide can be coupled to patient strategies to choose for another health system: an institutionalised exit (supported by governments and insurance schemes) is related to the public domain, a primary exit (having no insurance) is related to the private domain (Table 9.1).

The suggested typology can be interpreted further as follows: Citizenship is defined by the mechanisms of social coordination and the Patients' Rights Directive 2011/24/EU (Council of the European Union, 2011), as such citizenship is the core of EU cross-border health care. This also means that one should ascertain that the two archetypes (CBAS and TBAS) are different in kind by means of their cross-cultural and travel distance. It is now possible to reveal four types of patient mobility on the basis of the meta-model:

1. CBAS Public (Type 1): making public use of a proximal health system through a public insurance scheme, the patient is a citizen, in the framework of cross-border health care, which incorporates an institutionalised exit; other criteria are cultural proximity and access to information on behalf of the patient. As a case we could observe the Dutch crossing the border to receive health care in Belgium.
2. CBAS Private (Type 2): making private use of a proximal health system through their own financial means, the patient is a consumer, a medical tourist, which incorporates a primary exit; other criteria are cultural proximity and access to information on behalf of the patient. As a case we could observe patients from the UK opting to receive treatment in Belgium which is not covered by any health insurance scheme.[2]
3. TBAS Public (Type 3): making public use of a distant health system through a public insurance scheme, the patient is a citizen, in the framework of cross-border health care, which incorporates an institutionalised exit; other criteria are cultural distance and extensive seeking behaviour on behalf of the patient. As a case we could observe Arabic[3] patients who travel for cancer treatment to Germany with insurance coverage.[4]

4. TBAS Private (Type 4): making private use of a distant health system through their own financial means, the patient is a consumer, a medical tourist, which incorporates a primary exit; other criteria are cultural distance and extensive seeking behaviour on behalf of the patient. As a case we observe Russian[5] patients receiving stem cell treatment in Germany, paying out-of-pocket.

Concerning the other types of patient mobility – (1) the temporary movement of individuals (for example, tourists) who seek care outside their own country; (2) the permanent relocation of individuals who seek care outside their own country; (3) individuals living in border regions who cross borders largely as a matter of geographical convenience or cross-border care planning – it is not an issue: they fall easily in the public framework of cross-border health care. But from the moment patients start moving from a sending context (country A) to a receiving context (country B) with a primary reason to receive health care, which is not or not yet available in the sending context, and then a tension breaks out between what is considered as the public and the private domain. Therefore cultural and travel distance, incorporated by CBASs and TBASs, serve as a tool to conceptualise transnational health care. Finally, Laugesen and Bustamente's typology shows us the private and public divide, and its differences between the US and the EU. It is about having no insurance (private) (1-primary exit), or the exit is institutionalised and supported by governments and insurance schemes (2-institutional exit), which can be perfectly linked to the Glinos' typology (having cover/having no cover). Two other categories have been observed by Laugesen and Bustamente: searching for better and faster treatment abroad (private/public) (duplicative exit), having insurance coverage, but a service is not or partially covered (complementary exit). They are seen as less relevant at this moment for the typology, but it would be interesting for the development of the model to include them in the future. Especially complementary

Table 9.1 Typology Mainil et al.

Exit strategies of patient mobility (Laugesen and Vargas-Bustamente)	**Transnational settings (Mainil et al.)**		
	CBAS	**TBAS**	
Institutionalised exit	Public (Type 1)	Public (Type 3)	Cross-border health care citizenship (Carrera and Lunt)
Primary exit	Private (Type 2)	Private (Type 4)	Medical tourism consumerism (Carrera and Lunt)

exit – when only partial cover is available – is highly relevant in the light of the Patients' Rights Directive (co-payments): if the health care costs are higher than what the country of residence is providing to its international patients.

The typology makes a clear distinction between the public and private provision and use of health care services. This is necessary to keep the categories clear and transparent. However, in the clash between the Patients' Rights Directive and the mechanism of social coordination, one could observe a tendency which needs to be elaborated upon in the next part considering further developments. There are and will be cases in the future which are partially public and partially private in nature: if an international patient gets only partial coverage, and a part of the medical bill needs to be paid out of pocket, then the divide between citizenship and consumerism becomes even more complex and diverse. Our estimation of these developments can be observed in the next part.

Towards transnational patients and services?

As in Europe, patients themselves are becoming a mix of citizen and consumer, due to the recent European legislations; it becomes necessary and urgent to formulate in the first place considerations in the current European policy field: how willing are Europeans to travel abroad? What about socio-economic differences between the Member States? What about the economic climate in Europe? Secondly, some scenarios can be developed to describe how – while taking such considerations into account – the direction of European patient mobility will evolve. The four types of the typology are branded on these scenarios. Finally, challenges for health systems are linked with these scenarios.

Considerations regarding further developments (dynamics, drivers)

The typology introduced in the previous chapter can be linked with EU regulations. They offer also a framework for reflections about further developments. Major questions are as follows:

- What are the major motivations of EU citizens to travel for medical services across borders?
- How will the volume (financial volume, number of patients) of cross-border mobility develop, in total, under the regulations of the mechanism of social coordination, and under the mechanisms of the Patients' Rights Directive?
- Will there be differences between European countries?
- How will health systems adapt to these developments?
- What are positive developments, and what could be more critical issues?

For reflections on these questions, a couple of considerations should be taken into account. *A first consideration* concerns the theoretical willingness of the population to consume health services abroad. Approximately half of the EU population is in principle willing to travel abroad to receive medical services (Flash barometer, 2007). Major motivations are a hypothesised unavailability of necessary treatment, a perceived better quality and access to renowned specialists, quicker access and a cheaper treatment. At the same time, 42 per cent are not willing to travel. Major reasons are convenience, satisfaction with health care, language reasons, affordability, but also a lack in information. The non-willingness to travel abroad to receive medical treatment differs between EU Member States, due to the size of the countries, but especially also due to the socio-economic circumstances. In countries with a comparatively high socio-economic standard, citizens do not often see a reason to travel abroad. In countries with a comparatively low socio-economic standard, many citizens are discouraged by the affordability. There are also differences between socio-economic and cultural groups within countries. Younger people and people with higher education, inhabitants of urban areas and the self-employed are more often willing to travel abroad to receive medical services than others (Flash barometer, 2007).

A second consideration is also linked with socio-economic differences across the EU and addresses especially the differences in the costs of medical services with regard to third-party payers. They can also either be attracted by lower prices, disengaged or even overburdened by high prices in other countries.

The third consideration concerns the economic climate. The health policies within the Member States emphasise cost containment strategies. Private delivery of patient mobility or THC is much larger in other regions of the world (Asian and American continent), but it could also become larger in Europe, considering the stronger health system budgeting in the European crisis debate (Garel and Lombardi, 2011). Some services have been excluded from a reimbursement by social security systems, and also within social security systems, out-of-pocket payments have been raised. These developments might raise incentives for an extension of cross-border mobility for patients as well as for third-party payers. At the same time, health service providers, confronted with cost-containment strategies and limited budgets, might develop a strong interest in attracting patients from abroad to create additional revenue.

These considerations can be used to reflect upon the probability of four scenarios. In a *first scenario, based on the assumption that the current regulations will not create the dynamics for more cross-border mobility*, the status quo remains. The access to social security systems and health services will be preliminarily organised by and within EU Member States. That means that citizenship will dominate and in its material dimensions preliminarily be organised within EU Member States. The cross-border mobility of patients as well as the role of patients as consumers will stay limited. The same holds for cross-border cooperation between third-party payers and providers. Some contracts between third-party payers and health service providers will exist, but they will stay exceptions to the rule. In this scenario only the first type of the public CBASs citizen travel to a proximal country (type 1) will still be in place. It is the type which is the least difficult to execute for a patient and the oldest form of the phenomenon.

In a *second scenario, based on the assumption that the current regulations are implemented*, a strengthening of citizenship, based on an amalgamation of social rights and limited consumerism, could push cross-border mobility forward. Social security systems will be extended and include access to health services abroad. Quality and safety issues as well as payments will be based on international or EU agreements. But the organisation of social security systems and health systems stays mainly in the competencies of the EU Member States. In this scenario type 1 (CBASs/public/proximic/citizenship) and type 2 (TBAS/public/distant/citizenship) will tend to expand, based on the larger public and social security dimension.

In a *third scenario, based on the assumption that the implementation of current regulations will create a dynamic with impacts on citizens, patients, health care service providers and third-party payers*, the material dimensions of citizenship and health systems (patients, third-party payers, health services) become 'transnational'. They will not be primarily organised within the boundaries of Member States any more. Social rights are fully linked with EU citizenship, and third-party payers as well as health service providers operate and develop structures across borders. In this scenario all four types will evolve and expand. The public/private divide will probably not fully disappear, but a transnational system of social security will dominate. This scenario is the most beneficial for the development of patient mobility in Europe and beyond. This means if transnational health care is matured practice, the types of CBAS and TBAS could vaporise in one category or status of 'broad availability', generated from transnational networks and used by transnational patients.

The *fourth scenario* offers an alternative to the extension of citizenship by taking into account that also the consumer role of the patient could be strengthened, going hand in hand with a limitation of social rights. Health policies will emphasise cost containment strategies and private expenditures (private health insurances and out-of-pocket payments) will go hand in hand with deficits in the dimensions access, availability, affordability, quality and safety. Health policy makers will downsize the public character of health systems, strengthen private markets and accept cross-border supply of and demand for health services as private or market solutions. The patient acts (or is forced to act) as a *homo economicus*, looking for private solutions. Health service providers compete to attract patients from abroad. In this scenario type 2 (CBASs/private/proximic/consumerism) and type 4

(TBASs/private/distant/consumerism) will tend to become the main types, creating a departure from the European social security model.

The discussion of the scenarios can be combined with a reflection on the impact on the performance of health systems. The overarching goals of health systems have been defined as health, responsiveness and risk protection (and the instrumental goal efficiency). The main functions of health systems – steering, service provision, resource generation and financing – should contribute to the realisation of the overarching goals. An assessment of the consequences of cross-border mobility, addressing the overarching goals and main functions of health systems, can be organised around the Health System Performance Assessment (HSPA) dimensions.

With regard to the health of populations, cross-border mobility might have positive effects on the access to health services (at least for certain patient groups), either to closer services (border regions), specialised services (also with regard to rare diseases), to services of a better quality or to cheaper treatments (affordability). Positive impacts on the responsiveness (familiarity, meeting standards of perceived quality) can be expected, especially in a more competitive environment.

The continuity of the treatment (the management of interfaces in the process of treatment across borders), the assurance of safety and quality standards and liability rights might be challenging, but they are addressed by the Patients' Rights Directive and in principle manageable on the basis of EU-wide regulations. This holds also for the appropriateness of treatments. Major challenges are not so much based on medical services being delivered within the framework of regulations with regard to cross-border mobility. They are linked with services being offered beside the regulated system (for example, aesthetic surgery or services with questionable need or quality). A further challenge is linked with access to services which are not offered in a country due to political reasons (for example, pre-implantation genetic diagnosis (PID), abortion, euthanasia and so on).

The most relevant challenges concern challenges being linked with equity, fairness and sustainability. There is a high probability that the benefits and costs of cross-border mobility will not be equally distributed. It is not clear yet who will be the winners and the losers. Countries with a comparatively low GDP per head and low wages in the health sector might create additional revenues, also being beneficial for their balance of trade. On the other hand, the inflow of money might be counteracted by payments based on the Mechanisms of Social Coordination (social security mechanisms have to cover the prices of the country where services are delivered) or on the Patients' Rights Directive (especially wealthy patients decide to consume and pay for services abroad). Therefore, and in the light of EU health policy aims, objectives and values, it is a critical economic as well as ethical question whether cross-border mobility is mainly an opportunity for countries with a comparatively high socio-economic standard (cost containment at home and access and availability of health services abroad), while the stimulation of the development of the access to and availability of health services is not so favourable in countries with somewhat lower socio-economic standards.

In the light of citizenship versus consumer, the mechanisms of social coordination and the Patients' Rights Directive might develop as complementary regulations. But the Patients' Rights Directive strengthens also the consumer role. It can be linked with the political idea of a transformation of encompassing welfare states to three-tier systems, with a social security system covering basic services, private insurance covering additional services (like costs to services offered abroad and not fully covered under the Patients' Rights Directive) and a higher amount of out-of-pocket payments.

Concluding remarks

Taking into account the several mixed roles a patient can take, the construction of citizen's rights is determined by a divide between a conceptual difference in the roles of citizens and consumers, and the (political) reality: policies addressing citizen's rights can and have been linked with the idea to strengthen consumer roles. On the one hand this depends on the development of health policies in EU Member States, described in the four scenarios in this paper. On the other hand, European citizens, following the findings of the Eurobarometer, will take initiative in order to preserve their own well-being. If more conservative health policies will be the future in the EU, there will be more social spill-over effects. The (supra-)national governments will still have to take ad hoc decisions on the basis of a more decisive international patient. The EU Patients' Rights Directive offers a chance to European nation states to act more in favour of a European public health strategy. This means that patient mobility in Europe could be seen as a driver for change, both from a patient and from an institutional perspective. The constructed typology offers an understanding of the complex reality of patient mobility; furthermore it can be linked to the four scenarios. Globally a process of globalisation in health care has taken place. Transnational organisations in the form of insurance schemes or hospital chains are a given reality. A context-controlled steering mechanism such as the EU Patients' Rights Directive cannot stop this globalisation movement, but it can insert an equity balance for both national and international patients, if properly applied by the EU Member States. Therefore as a future research agenda, the monitoring of this application by the EU members onto the European Patients' Rights Directive, as well as the monitoring of transnational health care development, will become an important necessity.

Notes

1. Proximity or distance could even be translated as a criterion in relation to affordability: if there is distance between the budget of the patient and the price of

the treatment, this also means that financial proximity could be related to the EU directive on patient rights: (partial) coverage indicates proximity.
2. Developments in this type could be subject to the implementation of the EU Directive on patient rights.
3. Other nationalities are possible here. Arabic patients seem to comprise a large group of international patients in Germany (Anon. (2012). Kliniken umwerben Luxusklientel. Die Behandlung reicher ausländischer Patienten ist Lohnend. Stuttgarter Nachrichten 19/03/2012, 479 words).
4. The EU directive on patient rights and the mechanisms of social coordination will create new options for these patients.
5. Other nationalities are possible here. Russian patients seem to comprise a large group of international patients in Germany (Anon. (2012). Kliniken umwerben Luxusklientel. Die Behandlung reicher ausländischer Patienten ist Lohnend. Stuttgarter Nachrichten 19/03/2012, 479 words).

References

Bell, D., Holliday, R., Jones, M., Probyn, E. and Sanchez Taylor, J. (2011) 'Bikinis and Bandages: An itinerary for cosmetic surgery tourism' *Tourist Studies* 11, 2, 139–55.

Carrera, P. and Lunt, N. (2010) 'A European perspective on medical tourism: The need for a knowledge base' *International Journal of Health Services* 40, 3, 469–84

CMS Centers for Medicare and Medicaid Services (2010) *Medicare Coverage Outside of the United States* (Centers for Medicare and Medicaid Service).

Commission of the European Communities (2008) *Proposal for a Directive on the Application of Patients' Rights in Cross-Border Healthcare* (Commission of the European Communities).

Council of the European Union (2011) *Directive on Cross-Border Health Care Adopted* (Brussels, 7056/11, PRESSE 40).

European Economic Community (1957) *Treaty Establishing the European Economic Community* 25 March 1957, 298 UNTS 3.

Flash Barometer (2007) *Cross-Border Health Services in the EU – Analytical Report*. The Gallup Organization.

Garel, P. and Lombardi, G. (2011) 'The crisis, hospitals and healthcare' (European Hospital and Health Care Federation (HOPE)).

Glinos, I. A., Baeten, R., Helbe, M. and Maarse, H. (2010) 'A typology of cross-border patient mobility' *Health and Place* 16, 6, 1145–55.

Judgements Kohll and Decker (1998) 28 April 1998, cases C-120/95 and C-158/96.

Laugesen, M. J. and Vargas-Bustamante, A. (2010) 'A patient mobility framework that travels: European and United States-Mexican comparisons' *Health Policy* 97, 2–3, 225–31.

Legido-Quigley, H., Nolte, E., Green, J., la Parra, D. and McKee, M. (2012) 'The health care experiences of British pensioners migrating to Spain: A qualitative study' *Health Policy* 105, 1, 46–54.

Mainil, T., Van Loon, F., Dinnie, K., Botterill, D., Platenkamp, V. and Meulemans, H. (In Press) 'Transnational health care: towards a global terminology' *Health Policy* (Accepted for publication).

Official Journal of the European Union (1992) *Treaty on European Union (EU)*, 7 February 1992, OJ (C 191) 1, 31 ILM 253.

Official Journal of the European Union (2008) *Consolidated Version of the Treaty of European Union* (Brussels: Official Journal of the European Union).

Ormond, M. (2011) *International Medical Travel and the Politics of Therapeutic Place-Making in Malaysia*, PhD thesis (School of Geography and Geosciences, University of St. Andrews, UK).

Palm, W. and Glinos, I. (2010) Enabling patient mobility in the EU, in E. Mossialos, P. Govin, R. Baeten, T. K. Hervey (eds) *Health Systems Governance in Europe. The Role of European Union Law and Policy* (Cambridge: Cambridge University Press), 509–60.

Snyder, J., Crooks, V. A., Adams, K., Kingsbury, P. and Johnston, R. (2011) 'The "patient's physician one-step removed": the evolving roles of medical tourism facilitators' *Journal of Medical Ethics* 37, 9, 530–34.

Turner, L. (2011) 'Canadian medical tourism companies that have exited the marketplace: Content analysis of websites used to market transnational medical travel' *Global Health* 7, 1, 40.

Part III

Entanglements with Medical Tourism: Policy, Management and Business Responses

10
Canadian Medical Travel Companies and the Globalisation of Health Care

Leigh Turner

The key findings of this chapter are as follows:

- Though residents of Canada have access to medically necessary, publicly funded treatment available through provincial health care systems, some Canadians travel abroad and receive care at international medical facilities.
- Responding to increased public interest in medical travel and playing an important role in publicising global market for health services, many Canadian companies market transnational medical travel.
- Canadian medical travel companies advertise diagnostic imaging, orthopaedic surgery, bariatric surgery, cosmetic surgery, ophthalmologic surgery, in vitro fertilisation, dental care and other procedures.
- Some Canadian medical travel companies promote stem cell injections, 'Liberation therapy' for multiple sclerosis (MS), commercial surrogacy and additional interventions that are either illegal in Canada or clinically unproven and unavailable at Canadian hospitals and clinics.
- This chapter provides a comprehensive overview of Canada's medical travel industry.

Introduction

Canadians travel abroad for many different types of medical procedures. There are reports of Canadians travelling to international health care facilities for hip and knee replacements, bariatric surgery, ophthalmologic procedures, cosmetic surgery, dental care, cancer treatments, stem cell interventions, kidney transplants, commercial surrogacy, in vitro fertilisation and 'liberation treatment' for MS. Exactly how many Canadians go abroad every year for medical care is unknown; there are significant gaps in quantitative data concerning medical travel originating in Canada. However, sufficient evidence exists to have some insight into the types of procedures Canadians seek beyond Canada's borders. There are many newspaper reports describing journeys of individual medical travellers and some scholarly analyses of post-operative care required by Canadians after they received care outside Canada. In contrast, surprisingly little is known about the businesses that collectively constitute Canada's medical travel industry. Acknowledging several informative studies based upon interviews with Canadian medical tourism facilitators as well as content analysis of risk disclosure on websites of Canadian medical tourism companies, at present there is no comprehensive overview of Canada's medical travel industry (Johnston et al., 2011, p. 416; Penney et al., 2011, p. 17; Snyder et al., 2011, pp. 530–534). This chapter addresses the present gap in scholarship by providing an introduction to the current state of Canada's medical travel industry. Examining this subject reveals variation within Canada's medical travel industry and offers insight into how medical travel companies market transnational health care to prospective clients.

I began tracking and studying Canadian medical tourism companies in 2006 (Turner, 2007, pp. 73–77). I have followed the development of Canada's medical travel industry for 6 years. In addition to exploring ethical dimensions and policy implications of medical tourism, I have tracked individual companies with the intention of developing an empirical analysis of Canada's medical travel industry. Over the course of my research it became apparent that 'medical tourism companies' are an important but limited segment within a larger medical travel industry. Many Canadian companies promote health-related travel. However, not all of these businesses position themselves as 'medical tourism' facilitators. One goal of this chapter is to distinguish different types of companies from one another while also examining how these distinctive types of businesses collectively constitute Canada's medical travel industry.

Locating Canadian medical travel companies

I used multiple strategies when searching for Canadian medical travel companies. While numerous sources list some Canadian medical travel businesses, when I commenced my research there was no single resource containing a comprehensive record of all Canadian medical travel companies. I therefore used numerous search methods to build a database of Canadian medical travel companies. Using Google's search engine, I repeatedly conducted internet searches for Canadian medical tourism companies. Key terms used when conducting such searches included 'medical tourism Canada', 'medical tourism company Canada', 'medical tourism agency Canada' 'medical tourism facilitator Canada', 'medical tourism broker Canada', 'medical tourist Canada', and 'cross-border healthcare Canada'.

These terms were also used when searching for news media reports of Canadian medical travel companies. Digital archives of Google News Canada and ProQuest Newsstand were used to find news media reports describing Canadian medical tourists and the companies that organised their trips to international medical facilities. Supplementing search for archived newspaper articles, Google Alerts were used to provide prompt notification of publication of relevant news media reports. In addition, Industry Canada's website was a useful resource when trying to identify federally incorporated Canadian medical tourism companies. While few medical travel companies are listed in this database, it provided several useful leads. I was able to identify additional businesses by locating a short list of Canadian medical travel companies (Floyd et al., 2006). Attendance at a trade show on medical tourism held in Toronto in 2009 also provided me with insight into Canadian medical tourism companies seeking to market Indian health care facilities to prospective Canadian clients. Finally, participation in academic workshops and conferences held at the University of Toronto and Simon Fraser University provided me with an opportunity to meet several Canadian medical travel facilitators. These multiple search strategies complemented one another and facilitated development of a database of Canadian medical travel companies.

Using various search methods, in total I located 63 businesses that have a head office or affiliate office in Canada and market medical care provided at hospitals and clinics located outside Canada. Canada's medical travel industry is quite turbulent; as new companies are established other businesses leave the marketplace or merge with competitors (Turner, 2012, pp. 371–373). Following construction of the database, functioning medical travel companies were distinguished from businesses that had exited the marketplace. Elsewhere I examine the 28 Canadian medical travel companies that no longer market international health services to prospective customers (Turner, 2011, p. 40). Thirty-five companies remain operational and continue promoting medical travel to prospective clients. In this chapter I limit analysis to Canadian medical travel companies that remain in business.

I began searching for medical tourism companies with the assumption that businesses promoting transnational medical travel operate according to a common business model. Indeed, the first several companies that I found were all quite similar to one another and I did not notice significant differentiation within the medical travel industry. However, as my research progressed I began noticing important differences in the types of companies that market transnational and intranational health care. Companies engaged in promoting medical travel operate according to somewhat different models; they cannot all be neatly classified as medical tourism companies.

Canada's segmented medical travel industry

Of the 35 companies that are involved in coordinating medical travel and located in Canada, 18 firms fall within the category of what most health researchers would classify as medical tourism companies (Connell, 2006, pp. 1093–1100; Mason and Wright, 2011, pp. 163–177; Snyder et al., 2011). These companies promote health services provided at hospitals and clinics located in Barbados, Costa Rica, India, Mexico, Thailand and elsewhere; they also market holiday excursions and stays at hotels or resorts. Such businesses typically promote access to treatment at distant health care facilities, offer to arrange air travel and hotel accommodations, list different medical procedures, transfer medical records and promote 'all-inclusive' medical tourism packages (Mainil et al., 2011).

Seven additional companies promote cross-border travel to the US as well as private for-profit clinics within Canada. These businesses – similar in some respects to medical tourism companies – market medical travel but do not send their clients to distant international destinations. Existence of such companies is unsurprising. While it appears that on an annual basis only modest number of Canadians seek health care in the US, studies document some cross-border health-related travel from Canada to US medical facilities (Korcok, 1997, pp. 767–770; Katz et al., 1998, pp. 225–235, 2002, pp. 19–31; Eggertson, 2006, p. 1247). In general, these companies market prompt access to medical procedures available at facilities in the US rather than promoting health care combined with holidays.

In addition to identifying both medical tourism companies and cross-border medical travel facilitators, I found five businesses occupying narrowly defined niches within the medical travel industry. In particular, I identified two Canadian companies that advertise health services to individuals with MS and specialise in coordinating access to testing for Chronic Cerebrospinal Venous Insufficiency (CCSVI) and 'Liberation procedure'. Three additional companies advertise bariatric surgery and cosmetic surgery performed at health care facilities in Mexico.

Next, I discovered four businesses marketing private health insurance products that permit Canadian citizens to obtain access to health care in the US and Canada. These businesses offer critical illness insurance plans and also help their clients coordinate care outside Canada.

Finally, I found one Canadian company that markets medical procedures to uninsured and underinsured US citizens. Though located in Canada, this company does not market health services to Canadian citizens. Rather, taking advantage of Canada's proximity to the US as well as the approximately 50 million uninsured individuals in the US, it markets medical care to US citizens who must pay out of pocket for treatment.

I began my research by trying to locate and study Canadian medical tourism companies. In time, I found that Canada's medical travel industry is considerably more complicated than I had anticipated; not all medical travel companies are medical tourism facilitators. Medical travel companies can be sorted into different categories even though they all promote transnational health care. They occupy distinct market niches, promote different kinds of medical procedures and advertise different types of services to prospective clients.

Content analysis of company websites

Content analysis of company websites was used to gather data concerning key features of all identified medical travel companies (Pope et al., 2000, pp. 114–116). First, I recorded where medical travel companies are located within Canada. Second, websites were analysed for the purpose of establishing where Canadian medical travel companies offer to send their clients. In particular, I assembled data concerning destination nations and, where noted, specific health care facilities promoted by medical travel facilitators. Some company websites list a single destination for prospective customers. Other companies identify numerous possible destinations for health care. Third, I recorded what medical procedures and medical specialties businesses market. Some company websites list particular clinical specialties. Other company websites provide lengthy lists of medical procedures. Yet other websites list both clinical specialties and particular medical interventions. This topic was explored to determine the types of health services marketed by medical travel companies. Fourth, I extracted core marketing messages of medical travel companies and then summarised them in abbreviated form. Fifth, I analysed company websites to determine whether businesses offer to book travel, make hotel reservations and coordinate tours and side trips in destination nations. This subject was addressed to explore how many medical travel companies promote tourism and travel-related services in addition to marketing health care.

Medical tourism companies

In total, I found 18 businesses marketing health care provided at such international destinations as Costa Rica, India and Thailand (see Table 10.1). Most of these companies advertise a wide range of health services. Many of them market air travel, hotel accommodations and holiday excursions that can take place before or after receiving medical care. These companies take medical care and tourism and fuse them together to create a novel type of business enterprise.

Table 10.1 Canadian medical tourism companies

Company	Location	Destination countries	Health services marketed	Marketing message	Book travel	Book hotel	Book tours
Indus Health Tours	Vancouver, British Columbia; Tilak Nagar, India	India	Cardiology, orthopaedics, dental surgery, plastic and cosmetic surgery, minimally invasive surgery, ophthalmology, oncology, neurology, neural surgery, gastroenterology, urology, nephrology, gynaecology, Ayurveda, meditation, Yoga, spa	Access to affordable, timely, and high-quality health care	Yes	Yes	Yes
Meditours	Kelowna, British Columbia	India	Orthopaedic surgery for shoulder, hip, knee, elbow, and wrist, plastic surgery, 'liberation procedure', Thalamontomy treatment for Parkinson's, Stem cell treatment for hip necrosis, hip resurfacing, knee replacements, treatment for dystonia, tattoo removal	Access to affordable, timely, and high-quality health care in exotic settings	Yes	Yes	Yes
Metamorphosis Medical Retreats	Vancouver, British Columbia	Thailand	Plastic and cosmetic surgery, dental surgery, bariatric surgery, gender reassignment surgery, non-surgical interventions	Access to affordable and high-quality cosmetic surgery in exotic settings	Not noted	Yes	Yes

Passport Medical	Vancouver, British Columbia	Costa Rica, El Salvador, Mexico, Panama, South Africa	CCSVI Liberation Treatment, Dental procedures, Cosmetic surgery, Orthopaedic surgery, IVF fertility treatments, ophthalmology Diabetes treatment, weight loss surgery, IVF, IVF with Egg Donation abroad, Commercial surrogacy	Access to affordable, timely, and high-quality health care	No	Yes	Refer
Surgical Tourism Canada	Vancouver, British Columbia; Seattle, Washington	India, Mexico, US	CCSVI/MS Testing and Liberation Procedure, stem cell therapy, orthopaedics/spine surgery, bariatric surgery, cosmetic surgery, dental procedures, cardiology, oncology, IVF, neurosurgery, nephrology, private physician consultations/2nd opinions, preventative health checks, diagnostic services, MRI services, PET scan services, wellness clinics for complementary and Ayurvedic medicine	Access to affordable, timely, and high-quality health care	Yes	Yes	Not noted
Medical Concierge	Calgary, Alberta	US, Italy, Dubai (UAE), Canada, Singapore	Diagnostic testing, surgical procedures, spa care, 'posh prenatal' care, cataract surgery, other treatments	Access to timely and high-quality health care	Refer	Refer	Not noted

Table 10.1 (Continued)

Company	Location	Destination countries	Health services marketed	Marketing message	Book travel	Book hotel	Book tours
Overseas Medical Services Canada Inc.	Calgary, Alberta; Portland, Oregon; Loja, Ecuador; Saudi Arabia	Ecuador	Human Umbilical Cord Blood Cells, additional surgical procedures and therapies	Access to affordable, timely, and high-quality health care	Yes	Yes	Yes
Star Health Vacations	Edmonton, Alberta	India	Joint surgery, face treatments, breast surgery, body plastic surgery, skin treatment, male plastic surgery, plastic and reconstructive surgery, spa treatments, energy healing, hypnosis, gem therapy, guided imagery	Access to affordable and high-quality health care	Not noted	Yes	Yes
Surgical Escape	Calgary, Alberta	Costa Rica; Barbados; Mexico	Cosmetic and plastic surgery, bariatric surgery, orthopaedics, dentistry, ophthalmology, medical check-ups, dermatology, IVF	Access to affordable, timely, and high-quality health care	Yes	Yes	Yes
Global Healthcare Connections Inc.	Saskatoon, Saskatchewan	Canada, Costa Rica, Dominican Republic, India, Mexico, Singapore, Thailand, US	Dental, Cardiology, Bariatric surgery, Cosmetic non-surgical and surgical, Diagnostic Imaging, Hair transplant, MS Treatment, Oncology, Orthopaedics and Spine, Ophthalmology, Stem Cell and Regenerative Treatments, second opinions	Access to affordable, timely, and high-quality health care	Yes	Yes	Yes

Choice Medical Services	Winnipeg, Manitoba	Canada, Costa Rica, Cuba	Cosmetic and plastic surgery, diagnostic procedures, general procedures, women and men's health, vascular surgery, orthopaedic surgery, obesity management, eye surgery, nose and throat surgery, gynaecological surgery, neurosurgery, preventative check-up, internal medicine, dental procedures, neurology, cardiology	Access to affordable, timely, and high-quality health care	Yes	Yes	Yes
Aalpha International Medical Tourism Organisers Inc. (AIMTO)	Brampton, Ontario	India	Heart, dermatology, cosmetology and plastic surgery, cancer treatments-oncology, dental, ENT, nephrology and urology, neurology and neurosurgery, obstetrics, gynaecology and infertility treatments, ophthalmology, orthopaedics and joint replacement, paediatrics, physical medicine and rehabilitation, preventive health check-up, psychiatry, Ayurveda	Access to affordable, timely, and high-quality health care	Yes	Yes	Yes
CMN Inc. (Canadian Medical Network Inc.)	Thornhill, Ontario	Argentina, Brazil, Costa Rica, France, India, Malaysia, Poland, Singapore, South Africa, Thailand, Turkey, US	Gastric bypass, heart bypass, heart valve replacement, hip replacement, knee replacement, mastectomy, spinal fusion, additional tests and treatments	Access to affordable, timely, and high-quality health care	Yes	Yes	Not noted

Table 10.1 (Continued)

Company	Location	Destination countries	Health services marketed	Marketing message	Book travel	Book hotel	Book tours
MEDLINK GLOBAL INC.	Toronto, Ontario; Affiliates in UK, US and Pakistan	Thailand, Costa Rica, India, Singapore, Malaysia, Turkey	Joint replacement surgery, cosmetic surgery, prevention and alternative medicine, weight loss surgery, ophthalmic procedures, cardiac surgery, reconstructive surgery, diagnostic imaging	Access to affordable, timely, and high-quality health care	Yes	Yes	Yes
Debson Medical Tourism	Verdun, Quebec; New York, New York	Austria, Bahamas, Barbados, Belgium, Brazil, Bulgaria, Canada, Colombia, Costa Rica, Germany, Hungary, India, Israel, Italy, Jordan, Lebanon, Malaysia, Mexico, Panama, Peru, Singapore, South Africa, South Korea, Spain, Switzerland, Thailand, Turkey, Ukraine, the UK, US	Cancer, cosmetic, dental, eye, fertility, heart, orthopaedics and spine, stem cell, transplantation, weight loss, wellness and spa	Access to affordable, timely, and high-quality health care	Yes	Yes	Yes

GoSculptura, Inc	Montreal, Quebec	Argentina, Brazil, Columbia, Costa Rica, Dominican Republic, India, Mexico, Poland, Thailand	Plastic surgery, gastric bypass, gastric banding, sleeve gastrectomy, liposuction, cosmetic dentistry, hair transplantation, ophthalmology surgery, dermatology, IVF	Access to affordable and high-quality plastic surgery in exotic locations	Yes	Yes	Yes
Health Services Interna-tional/Services Sante International	Gabrielle d'Anneville, Quebec	Cuba	Retinitis pigmentosa, vitiligo and psoriasis, orthopaedic surgery, dental care, hair loss, diagnostics, cosmetic surgery, diabetes, detoxification, neurological rehabilitation	Access to timely and high-quality health care	Refer	Yes	Yes

Locations

Of the 18 medical tourism companies, 5 are based in British Columbia, 4 are located in Alberta, 1 is based in Saskatchewan, 1 is in Manitoba, 4 are in Ontario and 3 are based in Quebec.

Destinations

Seven of the companies marketed one country as a health care destination, two businesses marketed three health care destinations, three companies marketed five countries as destinations, one company marketed six destination nations, two companies marketed eight destination nations, one company marketed nine destination nations, one business marketed twelve destination nations and one business listed thirty potential destination nations. Combined, the 18 medical tourism companies list a total of 38 different destination nations. Eleven companies list India as a potential medical destination, nine companies list Coast Rica, seven companies list Thailand, seven list Mexico, six companies list the US, five companies list Singapore, five companies list Canada, three companies list South Africa, three companies list Brazil, three companies list Malaysia, three companies list Turkey, three companies list Barbados, two companies list Italy, two list Cuba, two list the Dominican Republic, two list Poland, two list Argentina, two list Panama, two list Colombia, two list Israel and two list the United Arab Emirates (with one company indicating Dubai as a potential destination and one company identifying Abu Dhabi as a possible destination site). El Salvador, the UK, Ukraine, Switzerland, Spain, South Korea, Ecuador, France, Austria, Bahamas, Belgium, Bulgaria, Germany, Hungary, Jordan, Lebanon and Peru are all listed by a single medical tourism company. The list of marketed destination nations, and the number of companies that mention them, reveals the variety of destinations marketed by medical tourism companies based in Canada as well as the most common destination nations promoted by Canadian medical tourism companies.

Advertised health services

Most Canadian medical tourism companies market comprehensive baskets of health services. However, some companies take niche positions by advertising restricted range of medical interventions. For analytic purposes, medical travel companies can be classified as 'generalist', 'specialist' and 'intermediate' firms. The former category includes businesses offering, for example, orthopaedic procedures, infertility treatments, cosmetic surgery, cardiac care, ophthalmology procedures, alternative medicine and dental surgery. A company limited to marketing cosmetic surgery procedures can serve as an example of a specialist medical tourism company. Of the 18 medical tourism companies promoting health care at global destinations, 14 businesses operate as generalist medical travel firms. Two companies are specialist firms. One business specialises in cosmetic surgery. The second firm predominantly markets procedures related to administration of human umbilical cord blood cells. Two businesses fall into neither of these categories and can be classified as having 'intermediate' marketing models. Of these latter businesses, one company emphasises in its marketing claims access to cosmetic surgery but its list of procedures includes cosmetic surgery, dental surgery, bariatric surgery, gender-reassignment surgery and non-surgical interventions. The other firm states that it primarily markets alternative health care but also advertises orthopaedic surgery and reconstructive surgery.

Marketing messages

Most medical tourism companies promoting travel to global destinations use core marketing messages that emphasise access to affordable, timely and high quality care. Sixteen medical tourism companies emphasise affordability of health care at international medical facilities. Access to timely health care is marketed by 15 companies. All 18 businesses market access to high-quality health care. Three businesses note that medical interventions can be obtained in exotic settings that appeal to tourists.

Travel, accommodations and tours

Medical tourism companies typically emphasise affordability of care, timely access to care and quality of care rather than tourism-related activities in their core marketing messages. Acknowledging that medical travel companies place greater emphasis on advertising medical interventions than promoting tourism, many businesses nonetheless market services associated with travel and tourism. Thirteen medical tourism companies offer to coordinate travel arrangements, 17 firms advertise the service of booking hotel reservations or otherwise arranging accommodations for clients, and 14 businesses offer to organise tours to local attractions located near where medical procedures are provided.

Companies marketing cross-border medical travel

Seven businesses market cross-border health care (see Table 10.2). These companies advertise medical procedures that can be obtained in the US and, in some cases, at private medical clinics in Canada. The US has a sizeable private health care sector in which it is possible to obtain ready access to care by paying out of pocket for treatment. Companies promoting cross-border medical travel as well as intranational travel within Canada market access to the large private health care sector in the US and the small but growing private health care sector in Canada (Silversides, 2008, pp. 1112–1113). In some respects they resemble the 18 businesses identified as medical tourism companies. However, cross-border medical travel companies differ in terms of where

Table 10.2 Cross-border medical travel companies

Company	Location	Destination countries	Health services marketed	Marketing message	Book travel	Book hotel	Book tours
OneWorld Medicare Inc.	Richmond, British Columbia	US, Canada	Hip replacements, knee replacements, MRI, CT, other procedures	Access to affordable, timely and high-quality care and medical access insurance	Yes	Yes	Not noted
Timely Medical Alternatives Inc.	Vancouver, British Columbia	Canada, US	MRI, CT Scans, PET Scans, Ultrasounds, Echocardiogram, colonoscopy, gastroscopy, orthopaedic surgery, neurosurgery, cardiac surgery, general surgery, knee replacement, gall bladder removal, angioplasty, cardiac bypass, arthroscopic shoulder surgery, spinal discectomy, weight loss surgery, hip replacement, cardiac ablation, additional diagnostic procedures and treatments	Access to affordable and timely health care	Not noted	Not noted	Not noted
Best Doctors Canada	Toronto, Ontario	Canada, US	Second opinions and review of treatment options, retesting of pathology, treatment at medical facilities in Canada and US	Access to affordable and high-quality care	Yes	Yes	Not noted

International Health Care Providers Inc.	Windsor, Ontario	US	Orthopaedic surgery, cardiology cancer, consultation services, psychiatric care, occupational ophthalmology, ear, nose and throat, neurology, endocrinology, paediatric endocrinology, gastroenterology, diagnostics, preventative diagnostics, bariatrics, gynaecology, vascular birthmarks, dermatology, sleep clinics, laparoscopic robotic prostate cancer surgery, CCSVI testing and treatment	Access to affordable, timely, and high-quality health care	Yes	Yes	Not noted
VIP Docs Inc.	Burlington, Ontario	Canada, US	MRI Scans, CAT Scans, PET Scans, CTA Scan, MRA Scan, MRCP Scan, TMJ MRI Scan, Ultrasound, Mammography, referrals to US physicians, surgery	Access to timely health care	Not noted	Refer	Not noted
VIP Health Options	Burlington, Ontario	US	MRI exams, Executive medicals, comprehensive health assessments, preventive health care, risk assessment, cancer care, diagnostic tests, critical illness insurance for out-of-country care, second opinions	Access to affordable, timely, and high quality health care; access to critical illness insurance	Yes	Not noted	Not noted
MedExtra	Saint-Laurent, Quebec	Canada, US	Diagnostic testing, second opinions, management of cancer treatments, minimally invasive surgery, major surgeries, specialised diagnostics	Access to affordable and timely care	Not noted	Not noted	Not noted

they propose sending their clients. They also differ in the extent to which they promote tourism- and travel-related activities in addition to marketing health services.

Locations

Of the seven cross-border medical travel companies, two businesses are located in British Columbia, four are based in Ontario and one is situated in Quebec.

Destinations

Six companies indicate that tests and procedures can be obtained in both the US and Canada; one business restricts itself to sending clients to US medical facilities. The lone company marketing medical travel primarily for diagnostic imaging is among the businesses offering access to health services in both the US and Canada.

Advertised health services

Of the seven companies marketing cross-border medical travel, four companies can be classified as generalist firms. Of the remaining three businesses, one company promotes access to many different kinds of care but places particular emphasis upon diagnostics and arranging second opinions, one company offers comprehensive services but emphasises preventive medicine and access to diagnostic tests and one business specialises in marketing diagnostic imaging.

Marketing messages

Six of the businesses marketing cross-border medical travel emphasise affordability of care. Six companies also market timely access to treatment. Four firms emphasise the high quality of care available at the destinations they promote.

Travel, accommodations and tours

Four of seven cross-border medical travel companies offer the service of booking travel and three offer to make hotel reservations. The possibility of booking tours to local holiday destinations is not addressed by these company websites.

Niche medical travel companies marketing 'liberation procedure' for MS

'Liberation therapy' and CCSVI as a diagnostic category for understanding MS receive extensive news media coverage in Canada. In response to public interest in 'liberation procedures', two medical travel companies in Canada dedicate themselves exclusively to promoting access to testing for CCSVI and 'liberation therapy' (see Table 10.3). These companies can be classified as niche medical travel firms.

Locations

These two businesses are based in Manitoba and Ontario.

Destinations

Of the two companies, one business advertises diagnostic imaging in the US and procedures in India. The other company markets both tests and procedures in India.

Advertised health services

Both businesses market testing for CCSVI and 'liberation procedure' for MS.

Marketing messages

Both companies emphasise affordability and high quality of marketed procedures; one business promotes timely access to treatment.

Travel, accommodations and tours

One company offers to book travel, both businesses offer to make hotel reservations and one advertises tours.

Niche medical travel companies marketing bariatric surgery and cosmetic surgery

Three companies market bariatric surgery performed at facilities based outside Canada (see Table 10.4). These companies are businesses that promote weight-loss strategies; they are not exclusively dedicated to promoting medical travel for bariatric surgery. However, since these businesses send their clients to Mexico, I have categorised them among the different types of companies promoting medical travel. In Canada, patients must often face long waits for bariatric surgery. These companies have responded to these delays by promoting prompt access to bariatric surgery at medical facilities beyond Canada's borders.

Locations

Of the three companies marketing bariatric surgery and cosmetic surgery, one business is located in Alberta and the remaining two companies are based in Saskatchewan.

Destinations

All three companies advertise surgery performed in Mexico.

Table 10.3 Companies marketing medical travel for 'CCSVI testing' and 'liberation procedure'

Company	Location	Destination countries	Health services marketed	Marketing message	Book travel	Book hotel	Book tours
CCSVI Clinic	Winnipeg, Manitoba	Canada, India, US	Doppler Ultrasound Screening for CCSVI; 'liberation procedure' 'stem cell therapy' in combination with venous angioplasty procedure	Access to affordable and high-quality testing for CCSVI and 'liberation therapy'	Yes	Yes	Yes
Liberation Gateway	Waterloo, Ontario; Detroit, Michigan	India	CCSVI Imaging and Stenting	Access to affordable, timely and high-quality 'liberation therapy'	No	Yes	Not noted

Table 10.4 Companies marketing medical travel for weight loss surgery

Company	Location	Destination countries	Health services marketed	Marketing message	Book travel	Book hotel	Book tours
Weight Loss For Eternity	Edmonton, Alberta	Mexico	Weight loss surgery, cosmetic surgery	Access to bariatric surgery and cosmetic surgery	Yes	Not noted	Not noted
Weight Loss Forever	Saskatoon, Saskatchewan	Mexico	Weight loss surgery, cosmetic surgery	Access to high-quality weight loss surgery and cosmetic surgery	Yes	Yes	Not noted
Weight No More Consulting	Saskatoon, Saskatchewan	Mexico	Weight loss surgery, cosmetic surgery	Access to high-quality bariatric surgery and cosmetic surgery	Yes	Yes	Not noted

Advertised health services

All of the companies marketing bariatric surgery at facilities outside Canada emphasise comprehensive access to weight-loss interventions. Their lists of specific procedures include cosmetic surgery.

Marketing messages

Their core marketing messages emphasise access to bariatric surgery and identify the variety of weight-loss programmes they offer.

Travel, accommodations and tours

All three businesses offer to book travel; two companies offer to make hotel reservations. None of the companies mentions organising tours and side trips before or after medical care is provided to clients.

Companies marketing health insurance products and assistance in coordinating medical travel

Four businesses market critical illness insurance products that help Canadians have the financial resources they need to obtain private health care within the US or at private clinics within Canada (see Table 10.5). Across Canadian provinces there are numerous restrictions on private health insurance for necessary medical interventions (Dhalla, 2007, pp. 89–96). These companies skirt these restrictions by providing insurance for health services delivered outside Canadian provincial health insurance plans. In addition to providing financial support for health care provided in the US or Canada's private health sector, these businesses help their clients arrange care at US medical facilities. In numerous respects these businesses differ from medical tourism companies. They do not, for example, market holiday excursions before or after treatment. Rather, they offer private health insurance products that can be used to cover expenses associated with obtaining out-of-pocket medical care.

Locations

Two of these companies are located in Alberta, one is based in Ontario and one is located in Quebec.

Destinations

In addition to offering insurance products these companies help their clients arrange care in the US.

Advertised health services

All four firms note that critical illness insurance can be used to access health care at US medical facilities. Three of the four businesses indicate that some types of tests and treatments are available at medical facilities in

Table 10.5 Companies marketing insurance products enabling access to care ir the US

Company	Location	Destination countries	Health services marketed	Marketing message	Book travel	Book hotel	Book tours
Acure Health Corp.	Calgary, Alberta	US	Critical illness insurance	Insurance program providing access to affordable, timely, and high-quality care	Yes	Not noted	Not noted
Canadian Equity Group Inc. sells and distributes MyCare Insurance Program	Calgary, Alberta	US	Health insurance providing access to medical reviews, diagnostic testing, access to treatment, and consultations provided by Mayo Clinic physicians	Insurance program providing access to affordable, timely, and high-quality care	Not noted	Not noted	Not noted
Right Choice Insurance Inc.	Toronto, Ontario	Canada, US	Critical illness insurance provides cash payment, access to second opinions, and access to out-of-province and out-of country care	Critical illness insurance providing access to affordable, timely, and high-quality health care in the US and select Canadian facilities	Not noted	Not noted	Not noted
Etfs Travel and Healthcare Solutions	Sherbrooke, Quebec	Canada, US	Health insurance product is intended to cover heart surgery, cancer, digestive disorders, ear, nose and throat, endocrinology, gynaecology, neurology, ophthalmology, orthopaedics, paediatrics, respiratory and urology	Private health insurance permitting access to timely and high-quality health care in the US and select Canadian medical facilities	Not noted	Not noted	Not noted

Canada. One insurance programme is restricted to enabling access to tests and procedures at three Mayo Clinic sites based in the US.

Marketing messages

Three of the four companies emphasise affordable access to care. All four companies promote timely access to health care and access to treatment at high-quality health care facilities.

Travel, accommodations and tours

Of the four companies marketing insurance products enabling access to care in the US as well as private clinics in Canada just one business clearly indicates that it coordinates travel for clients. None of the companies state whether they book hotel reservations and organise tours. These companies emphasise offering insurance products and ensuring that their clients can obtain timely access to care in the US or Canada. They help arrange care at US facilities but they do not organise logistics to the same extent as medical tourism companies and regional, cross-border medical travel companies. In addition, they do not promote tours and holiday excursions.

Medical travel company marketing health services to US citizens

One medical travel company based in Canada markets health services primarily to US citizens (see Table 10.6). The US has a large population of uninsured and underinsured individuals. This company, a companion firm to a business that markets medical travel for Canadians, promotes to uninsured and underinsured Americans affordable access to health care within eight US states.

Location

This firm is located in British Columbia.

Destinations

This business markets access to medical procedures available in eight US states as well as facilities in Canada.

Advertised health services

The company promotes such health services as cardiac surgery, general surgery, orthopaedic surgery, neurosurgery, High Intensity Focused Ultrasound for prostate cancer and additional medical interventions.

Marketing messages

This company emphasises access to affordable, timely and high-quality medical interventions.

Table 10.6 Medical travel company marketing health services to US citizens

Company	Location	Destination countries	Health services marketed	Marketing message	Book travel	Book hotel	Book tours
North American Surgery Inc.	Vancouver, British Columbia	Canada, US	Cardiac surgery, general surgery, orthopaedic surgery, neurosurgery, women's procedures, spinal surgery, High Intensity Focused Ultrasound (HIFU) treatment for prostate cancer, additional treatments	Access to affordable, timely, and high-quality health care	Not noted	Not noted	Not noted

Travel, accommodations and tours

The company does not indicate whether it books travel, makes hotel reservations or organises tours and other holiday excursions. It facilitates medical procedures and does not advertise additional services.

Canadian medical travel companies

Canada now has many businesses marketing access to medical procedures that are paid for out of pocket and provided at international medical facilities. Compared to the US and many countries in Asia, Canada does not have a large private health care sector (Steinbrook, 2006, pp. 1661–1664). Provincial health care systems provide universal access to 'medically necessary' health services; legislation restricts the types of medical procedures for which private clinics are allowed to charge patients (Flood and Archibald, 2001, pp. 825–830). The modest scale of Canada's for-profit, private health care sector appears to have resulted in the emergence of companies that serve as bridges or intermediaries to hospitals and clinics located outside Canada. Some of these businesses also market intranational travel to private, for-profit medical clinics located within Canada. Various insurance products provide Canadians with tools for gaining access to health care facilities in the US.

Presumably wanting to appeal to the largest possible number of prospective clients, most medical travel companies offer many different kinds of medical procedures. Some medical travel companies respond to treatment delays in Canada by marketing swift access to private health care at international facilities (Christou and Efthimiou, 2009, pp. 229–234; Legare et al., 2010, E17–21). Orthopaedic procedures, ophthalmologic procedures, bariatric surgery and other interventions that sometimes require lengthy waits for care in Canada are readily available within the private health care sectors of other countries. Many Canadian medical tourism companies sending their clients to such countries as India and Thailand, as well as medical travel companies promoting cross-border travel to the US, in their marketing messages emphasise timely access to care. Claims about offering access to high-quality care are also widespread. Clients of medical travel companies presumably want not just fast access to treatment but also access to professional medical care. Company websites offer messages intended to reassure clients about treatment available at international hospitals and clinics.

There are some circumstances where provincial health insurance plans in Canada permit reimbursement of expenses incurred as a result of receiving out-of-country medical care. In Canadian provinces, coverage of out-of-country care for elective procedures typically requires, at minimum, recommendation by a Canadian physician and pre-approval by provincial health ministries (Lindberg and Risk, 2007, pp. 1–18). In most instances individuals travelling abroad for elective medical procedures must pay out of pocket for treatment. Perhaps for this reason, many medical travel companies in Canada emphasise affordability of the procedures they market.

Some companies specialise in offering medical interventions that have not undergone clinical trials or regulatory review and are not approved for patient care in Canada. For example, medical travel companies marketing access to stem cell injections and 'Liberation Therapy' help their clients gain access to procedures that patients cannot obtain in Canada. From one perspective, offering such services promotes 'consumer choice' by giving clients access to procedures they wish to undergo. However, 'expansion of choice' exposes medical travellers to considerable risk by enabling them to gain access to medical interventions with unknown safety and efficacy profiles. To date, at least two Canadians with MS are reported to have died while undergoing 'Liberation Therapy' at medical facilities located outside Canada (Morrow, 2010; McClure, 2011).

This chapter does not examine all of the companies that Canadians can consider when deciding whether to seek care outside Canada. Canadians considering going abroad for treatment can also contact destination hospitals or select health care packages provided by medical travel companies based outside Canada. This chapter analyses medical travel companies based in Canada; it does not examine medical travel businesses and destination medical facilities located beyond Canada's borders.

Conclusion

- Canada's publicly funded provincial health care systems provide access to medically necessary care. Given that residents of Canada have universal access to medically necessary health services, individuals unfamiliar with health care in Canada might assume that few medical travel companies are likely to emerge in Canada. To the contrary, 35 businesses market medical travel.
- This chapter provides an overview of Canada's medical travel industry. Content analysis of company websites reveals where these businesses are based, the health care destinations they promote to prospective clients, the health services they market, their core marketing messages and whether companies market travel, hotel reservations and tours in addition to advertising health services.
- Acknowledging the possibility of a gap between how medical travel is marketed and behaviour of medical travellers, analysing websites of Canadian medical travel companies provides insights into how these businesses promote health services to prospective clients.

Medical tourism companies based in Canada have attracted some attention from researchers but there is limited analysis of company websites and little exploration of the many medical travel companies that do not neatly fit into the category of medical tourism businesses (Johnston et al., 2011, p. 46; Snyder et al., 2011). This chapter examines the medical travel industry of a country with a publicly funded system of universal access to health care. There are presumably some important differences between why Canadians go abroad for health care and why residents of other countries travel for medical interventions. Scholarly research examining medical travel will benefit from analysis of many different social, economic and cultural contexts.

To date, researchers addressing the phenomenon of 'medical tourism' appear particularly interested in studying companies that promote medical travel to India, Thailand and other distant, 'global' health care destinations. This chapter identifies and describes differences among different types of companies in Canada's medical travel industry. Some companies market health services available in countries located far from Canada. Other companies promote cross-border travel to the US as well as intranational travel within Canada. Yet other businesses promote specific interventions such as bariatric surgery and 'liberation procedure'. Some companies market insurance products and help Canadian clients obtain health care in the US and private facilities in Canada. Studies that focus strictly on medical tourism companies sending clients to distant nations risk missing or neglecting many of the businesses identified in this study. Businesses promoting medical travel are not reducible to medical tourism companies marketing health services and holidays available in faraway international locations. There is something to be gained from identifying differences within the medical travel industry and expanding the analytic lens to include both medical tourism companies and related businesses that promote medical travel but do not necessarily fit the standard model of medical tourism companies. Businesses promoting cross-border care, weight loss, 'liberation therapy' and private insurance coupled with assistance arranging health care in the US all market access to various types of medical travel.

At present there is no resource that health researchers, clinicians, policy makers, journalists and other individuals can use as an accessible guide to analysing both the general terrain of Canada's medical travel industry and the particular features of specific businesses. This chapter aims to contribute to scholarship by putting a guide into the hands of health researchers and other parties interested in further exploring territory that is fascinating, complex and raises numerous ethical, social, legal and public health concerns.

Acknowledgements

This chapter is a revised version of an article I published in *Globalization and Health*. I wish to thank BioMed Central and the editors of *Globalization and Health* for permission to reuse material from that publication.

References

Christou, N. and Efthimiou, E. (2009) 'Bariatric surgery waiting times in Canada' *Canadian Journal of Surgery* 52, 229–34.

Connell, J. (2006) 'Medical tourism: Sea, sun, sand and ... surgery' *Tourism Management* 27, 1093–100.

Dhalla, I. (2007) 'Private health insurance: An international overview and considerations for Canada' *Healthcare Quarterly* 10, 89–96.

Eggertson, L. (2006) 'Wait-list weary Canadians seek treatment abroad' *Canadian Medical Association Journal* 174, 1247.

Flood, C. and Archibald, T. (2001) 'The illegality of private health care in Canada' *Canadian Medical Association Journal* 164, 825–30.

Floyd, M., Izenberg, D., Kelly, B., Shimo, A. and Treble, P. (2006) 'Medical services directory' *Maclean's,* 25 April.

Johnston, R., Crooks, V., Adams, K., Snyder, J. and Kingsbury, P. (2011) 'An industry perspective on canadian patients' involvement in medical tourism: Implications for public health' *BMC Public Health* 11, 416.

Katz, S., Verrilli, D. and Barer, M. (1998) 'Canadians' use of US medical services' *Health Affairs* 17, 225–35.

Katz, S., Cardiff, K., Pascali, M., Barer, M. and Evans, R. (2002) 'Phantoms in the snow: Canadians' use of health care services in the United States' *Health Affairs* 21, 19–31.

Korcok, M. (1997) 'Excess demand meets excess supply as referral companies link Canadian patients, US hospitals' *Canadian Medical Association Journal* 157, 767–70.

Legare, J., Li, D. and Buth, K. (2010) 'How established wait time benchmarks significantly underestimate total wait times for cardiac surgery' *Canadian Journal of Cardiology* 26, E17–21.

Lindberg, M. and Risk, J. (2007) 'External review of the Ontario health insurance plan's out-of-country program' 1–18 July.

Mainil, T., Platenkamp, V. and Meulemans, H. (2011) 'Diving into contexts of in-between worlds: World making in medical tourism' *Tourism Analysis* 15, 743–54.

Mason, A. and Wright, K. B. (2011) 'Framing medical tourism: An examination of appeal, risk, convalescence, accreditation, and interactivity in medical tourism web sites' *Journal of Health Communication* 16, 163–77.

McClure, M. (2011) 'Woman with MS dies after treatment' *Winnipeg Free Press* 9 July 2011.

Morrow, A. (2010) 'Man dies after controversial MS treatment, doctor says' *The Globe and Mail* 18 November 2010.

Penney, K., Snyder, J., Crooks, V. and Johnston, R. (2011) 'Risk communication and informed consent in the medical tourism industry: a thematic content analysis of Canadian broker websites' *BMC Medical Ethics* 12, 17.

Pope, C., Ziebland, S. and Mays, N. (2000) 'Qualitative research in health care: Analysing qualitative data' *British Medical Journal* 320, 114–6.

Silversides, A. (2008) 'Canada Health Act breaches are being ignored, pro-medicare groups charge' *Canadian Medical Association Journal* 179, 1112–3.

Snyder, J., Crooks, V., Adams, K., Kingsbury, P. and Johnston, R. (2011) 'The patient's physician one-step removed: The evolving roles of medical tourism facilitators' *Journal of Medical Ethics* 37, 530–4.

Steinbrook, R. (2006) 'Private health care in Canada' *New England Journal of Medicine* 354, 1661–4.

Turner, L. (2007) 'Medical tourism: Family medicine and international health-related travel' *Canadian Family Physician* 53, 1639–41.

Turner, L. (2011) 'Canadian medical tourism companies that have exited the marketplace: Content analysis of websites used to market transnational medical travel' *Globalization and Health* 7, 1, 40.
Turner, L. (2012) 'Canada's turbulent medical tourism industry' *Canadian Family Physician* 58, 371–3.

11
The Ethical Management of Medical Tourism

Guido Pennings

The arguments of this chapter are as follows:

- More and more reliable data are needed to be able to construct a complete picture of the phenomenon of cross-border medical care.
- Medical tourism is the result of the liberalisation and commercialisation of health care.
- The ethical management of medical travelling should be organised on the basis of the benchmarks of fairness. This list of criteria avoids a too simplistic approach of the developments.
- The effects of medical travelling will be evaluated differently depending on the theory of just distribution of scarce resources: utilitarianism, egalitarianism, prioritarianism and sufficientarianism.
- The migration of medical personnel (the 'brain drain') is a perfect topic to study the effects of medical travelling and the implications of the theories of distributive justice.
- The commercialisation of health care invites new players to the field (private insurance companies, brokers, corporate hospitals and so on) whose rights and duties still have to be determined.

Introduction

Medical travelling or patient mobility is a phenomenon that contains many different types of patients and medical interventions. We will use the term to indicate people who purposefully leave their country to obtain medical treatment abroad. However, we will only consider 'standard' medical treatment and leave out specific types of procedures such as infertility treatment, organ transplantation and stem cell therapy (covered by other chapters), because they raise special ethical and management issues.

A crucial element to evaluate a phenomenon like medical travelling is evidence. At the moment, there is no generally accepted definition of 'medical traveller' or 'medical tourist'. This explains, at least partially, the high variations in guesstimates that can be found in the literature. McKinsey states that approximately 50,000 Americans go overseas while Deloitte estimates the number at 6 million in 2010, growing to almost 16 million in 2017. An analysis of the source of the numbers corroborates the suspicion that numbers are used as market tools (Youngman, 2009). It is obvious that many parties have an interest in presenting the movements as an established fact (hoping on a self-fulfilling prophecy) taking on huge proportions. Meanwhile, no one doubts the fact that the flows are steadily increasing.

Beside the lack of data on the phenomenon itself, there is also paucity of data regarding the real effects of medical travelling (MT) on the health care systems. We urgently need systematic and reliable empirical data to determine the impact of the flows on the health care systems, both in the home and in the destination countries. Collecting evidence on the impact of the patient flows is not only a huge task because of the complexity of the data but also because of the complexity of the existing structures and the lack of registration of such data in many developing countries.

The status of health care

Access to health care is a human right. As a consequence, governments have the task to guarantee this right. Social security systems are based on solidarity, collective responsibility and equal contributions in order to ensure accessibility of high quality care. Health care is considered as a special good because health is believed to be an essential precondition for well-being; a person can only fully enjoy the meaningful activities in life if he or she possesses a certain degree of health. An unhealthy person or a person with a disease or disability no longer possesses the full range of opportunities that people normally possess. Health care is meant to restore normal species functioning (Daniels, 1985).

The evolution of trade in health care is a consequence of the liberal ideology. On a worldwide scale, the General Agreement on Trade in Services (GATS) negotiations intend to liberalise all services. For a number of people, some activities should not be included in these agreements because they have a special status (Labonte, 2004). Privatisation of essential public services such as health care, education and drinking water may endanger equity and access to services by the most needy groups in society (UNCTAD Secretariat, 1998). From the moment that health care is tradable, it is subjected to the rules of the market and equitable access can no longer be guaranteed. Trade in health care services goes against the core duty of the state to provide basic health care. Although it is not necessary for the fulfilment of this duty that health care is in public hands, it is certain that extensive state regulation and control is needed. The provision of health services cannot be completely left open to free competition because it would automatically imply a serious set-back of the right to access. This liberalisation and commercialisation reaches its final expression in global medical travelling. 'Medical tourism' reveals the shape that medicine takes when it is commodified, subjected to international competition, and subsumed within a global market economy' (Turner, 2010).

This evaluation is crucial. Agencies like the World Bank and the International Monetary Fund, supported by the negotiations of the GATS, all push in the direction of privatisation and commercialisation of health care. However, by transforming health care into a service like any other, one implicitly or explicitly alters the status of health as a primary good. Private markets have different goals from public health care systems. The main difference is related to equity.

Evaluating medical travelling

How are we to evaluate medical travelling (MT)? In an increasing number of countries, MT takes on the proportion of a health care system reform. Such reforms can be judged by numerous criteria but the three core values are equity, efficiency and accountability. These goals have been further developed into the 'benchmarks of fairness' (Daniels et al., 2000). Five benchmarks express aspects of equity: the exposure of people to public health risks and to inequalities in the distribution of the social determinants of health, financial and non-financial barriers to equitable access to care; inequalities in the benefits for different groups; and the burden of health care cost among the sick and poor (Daniels, 2006). Two benchmarks elaborate on efficiency by highlighting administrative efficiency and clinical efficiency. The final benchmarks regard democratic empowerment and provider and patient autonomy as aspects of accountability. The ethical question from a broader perspective is how MT affects the benchmarks in the home and destination countries. More specifically, this chapter will address the question whether the flow of foreign patients to developing countries improves or worsens the existing health care system in terms of these benchmarks. Since this is a multi-criteria model, it is highly likely that trade-offs will have to be made. An intervention or measure may improve one aspect and simultaneously hamper another. This balancing will have to be done in a case-by-case evaluation, taking into account the local circumstances. This implies that a global evaluation of medical travelling as a worldwide phenomenon may not be very informative about the effects or characteristics in a specific country.

Travelling takes place between two countries, that is, the home country and the destination country. It is very well possible, and even likely, that the same patient flow is evaluated differently depending on the perspective. Countries with a structural deficit in their health care system may try to reduce the shortage by promoting travelling abroad. In Ireland, undue delays occur for most procedures (Healy, 2009). In the Netherlands waiting lists can be long, while Belgian waiting lists are usually negligible. This has become a significant cause of cross-border health care (Hermans, 2008). Organised projects to promote cross-border travel when health care systems are found lacking have been reported in Ireland, the UK, Norway, the Netherlands, Germany and Denmark (Glinos et al., 2010). In these circumstances, patient autonomy has increased in the home country since patients are now able to access care without extra costs and without long waiting times. Simultaneously, patient autonomy in the destination country may actually be less respected since these patients may have to wait longer due to a shortage of doctors and/or may no longer be able to pay for the treatment because the foreign patients caused inflated prices.

Each benchmark should be applied to MT. For instance, does the development of MT reduce access to public health? The World Bank and the International Monetary Fund have forced developing countries to cut public spending on education, child care, social welfare and health care (Meghani, 2011). Data on the evolution of the public health sector show that since the liberalisation of the health care sector, access to health care has worsened. Thus, not MT itself but the steps enabling MT to develop have had a detrimental effect on access to public health care. However, the same plan of liberalisation and privatisation of health care also channels money from the government to private clinics through tax breaks, lower import duties and so on (Meghani, 2011). Indirectly, the tax payers cover part of the bill of the medical tourist travelling to their country. Moreover, that money cannot be spent on public health care for their own citizens. Nevertheless, it will be very hard to show that certain effects are a direct consequence of the development of MT. MT may, for instance, reinforce a dual system in which rich nationals and foreigners have access to high-quality health care while the local poor have to be content with low-quality care in the public sector. This is the idea expressed in the image of 'islands of excellence in a sea of medical neglect' (Chinai and Goswani, 2005). Although a mixed system does not have to be problematic, it may undermine the public tier both economically and politically (Daniels et al., 2000). So, even if certain detrimental evolutions may not be caused by MT itself, MT may contribute to these evolutions and may stimulate or support the general ideology leading to these harmful effects (Ormond, 2011).

Fixing the reference point to determine whether or not MT leads to deterioration of the health care system in a country is crucial for the ethical evaluation. Since many people object to the current evolution and to MT in general, one tends to blame the medical tourism industry for all the deficits of the health care provision in poor countries. However, the present system already fails on numerous points: inequity, lack of access and low-quality care exist at present. Most developing countries spend around one per cent of their gross national product on public health care. It could be argued that liberalisation has worsened the situation, but this would have to be tested in specific country contexts. One frequently mentioned consequence of MT would be diminished access. However, whether or not this is true will depend to a large extent on the type of treatment. Many patients travel to poor resource countries for high technological interventions such as heart surgery. Access to these high-tech interventions will not change for the poor since they are not available to poor residents in the first place; they could never afford such treatments. The effect of MT on access, if any, would be limited to the middle and higher-middle class in the destination countries.

Theories of distributive justice

The ethical theory that one adopts is crucial for the evaluation of the effects of MT. Generally speaking, four theories on the just distribution of scarce resources can be distinguished: utilitarianism, egalitarianism, prioritarianism and sufficientarianism. Utilitarians will distribute the scarce resources so that aggregative well-being is maximised. Egalitarians aim at equality in opportunities, capabilities or well-being. Prioritarianists focus on those who are worst off; they apply Rawls' difference principle by which those who have the least should receive the most. Finally, sufficientarianists try to ensure that people do not fall below a specific threshold, frequently placed at the level of 'reasonably good health' (Cohen, 2011).

Our discussion focuses mainly on the prioritarianist position since this is the main concern expressed in the literature (Pennings, 2007). MT should not make those who are already worst off (poor patients in developing countries without access to basic health care) even worse off. However, to realise the impact of the different theoretical positions, let us consider the following situation. Imagine that MT improves the health of the travelling patients considerably while it lowers the health of the local patients to a much lesser extent. This situation would be acceptable for a utilitarian. It would also be acceptable for a sufficientarianist on the condition that the health of local patients remains above the threshold. However, it violates the egalitarian principle because it increases health inequalities (at least between the visiting patients and the local patients) and the prioritarian rule since the worst off are even worse off. If, on the contrary, the health care of the local patients would improve due to MT, but considerably less than that for visiting patients, only the egalitarian would object. However, even this finding depends on which groups of patients one compares: the health inequalities between the patients in the home country who can afford to have treatment at home and the travelling patients diminish and the inequality between the local patients (rich and poor) also diminishes. Some people conclude that, since the local patients also benefit, we can reasonably accept such a policy as long as the health of all groups improves (Mechanic, 2002). Others, however, will insist on reducing health inequalities by levelling up the health of the worst off first and foremost. The question then becomes whether there exists a policy that either increases the health of the local patients more or results in less inequality. It is not too difficult to see that this is obviously the case: already now regulations are proposed to make sure that local patients benefit more directly (see further). However, as for other poverty-reducing measures, it is also clear that more could be done.

The ethical evaluation is further complicated by the debate on how we are to determine harm. We need a definition or test in order to decide when people are harmed by a health care reform (or any other intervention for that matter). Feinberg (1986) distinguishes two tests: (1) the 'worsening' test – the intervention puts the people in a worse condition than they were before the intervention and (2) the 'counterfactual' test – the intervention puts the people in a worse condition than they would be in if the intervention had not been performed. The 'counterfactual' test converges with Morreim's proposal to take as a reference point to decide about harm not what will be the case but what should have been the case. 'A person is harmed at time T in respect R if his condition R is worse than it should have been at time T' (Morreim, 1983). The normative rule (which tells us what should have been) is that basic health care should be available at home. So, there are two ways to evaluate the effects of MT in the home country: firstly, by comparing with the situation as it is now (with the structural deficit in the health care system) or secondly, by comparing with the situation as it should be (decently developed health care system with equitable access to reasonable care). These two tests result in contradictory conclusions: compared to the present situation, patient autonomy has increased when they can go abroad; compared to the normative situation, their autonomy has decreased (be it less than when they would not be able to cross borders) since they have to suffer the social, financial and psychological inconveniences of travelling abroad. It seems that a country that sends or indirectly pushes its citizens abroad for health care that should be available at home is opting for second best: sacrificing patient autonomy for cost reduction. From an ethical point of view, this is hard to justify when the health care service patients are seeking abroad is part of what society has agreed to include in reasonable health care.

Unfortunately, it is too easy to treat international disparities in health as if they were national disparities. Health care is organised at the level of nations. It is unclear at the moment what rich nations should do to improve the health of the population of less wealthy countries (Daniels, 2006). It could be argued that every state above a certain welfare level has a humanitarian obligation to help people in grave need, regardless of what it does itself. This underlies the duty to provide general developmental aid. However, there is a stronger obligation when the consequences in another state follow from the behaviour of their own citizens in response to a state policy (Pogge, 2005;

Cohen, 2011). But suppose that the destination country itself is not doing much to improve the health of its citizens, is the health inequality then a matter of international justice? We would, at least for the time being, support a 'minimalist' view in which states have a negative duty to prevent harm to others, even if they do not have a positive duty to help the population in the other country. If the situation in the destination country gets worse because a policy in the home country foreseeably makes the health of the poor people in those countries worse, then this harm is unjust (Pogge, 2005). There are two possibilities: the home state actively promotes MT of its citizens by offering all kind of benefits (as was considered in West Virginia) or the home state indirectly promotes MT by its failure to secure universal health care or at least reasonable access to health care. So, if the home country causes travelling by its citizens by reducing public spending in health care at home, then it has a duty to do something about it when evidence shows that this deteriorates the situation of the poor in the destination country.

Brain drain

The brain drain is an interesting topic to be tackled by the benchmarks of fairness systems. It is common knowledge that developed countries such as Canada, the US and the UK have been (and still are) massively recruiting medical personnel from developing countries such as South Africa and the Philippines. These countries have been fighting the migration of medical personnel for decades because they are confronted with severe shortages at home. Both the Commonwealth and the World Health Organisation have issued declarations intended to discourage the targeted recruitment of health workers from countries which are themselves experiencing difficulties (Commonwealth, 2003; World Health Organisation, 2010). The shortage of health care personnel does not only exist at the international level between developing to developed countries; medical personnel also move from rural to urban regions and from public to private sector within a country. The 'global health conveyor belt' moves between different levels. How does MT affect these movements? It is important to keep in mind that MT of patients from rich countries equals the export of medical services for the destination country. In other words, it causes an internal brain drain: although medical personnel do not physically leave the country, they serve foreign patients. From the few countries that have data on this particular point, we know that there is a flow of medical personnel from rural to urban areas where the clinics for foreigners are located (Smith et al., 2009). Moreover, clinics serving foreign patients are all private (and frequently owned by foreign companies), thus causing a double drain from rural to urban and from public to private clinics. To a large extent, the bad effects of the brain drain are similar when the movements take place internally and externally. Although recommendations have been made by international organisations to remedy the migration of medically qualified people, these recommendations have not been extended to cover cross-border patient mobility from rich countries. Both the lack of visibility of MT (as mentioned, no one really knows how many patients travel, where and for what) and the failure to equate incoming patients with outgoing medical personnel may be responsible for this shortcoming.

Taking into account the rights of the patients, the first step to avoid the problem in both countries is that the home countries invest much more in their own health care system to make it more attractive so that more of its own citizens choose a health care profession. This would not only diminish the need to recruit qualified professionals from abroad but may simultaneously decrease cross-border travelling by patients. A second step would be to apply the compensatory measures proposed for countries recruiting foreign personnel equally to countries whose citizens travel abroad to receive medical care. The home countries should, for instance, support training in the destination countries, increase capacity building, facilitate transfer of technology, skills, and financial assistance. Although the codes of practice seem to use the idea of reciprocation, restitution or mutual benefit to structure the duties of the home countries, it seems highly unlikely (given the power imbalance) that the benefits will be split equally. Interestingly, two strategies are possible: either the home country invests money to increase its health care capacity or the home country contributes to the health care system of the destination country to increase its capacity. If the contribution merely compensates the amount of health care that is diverted to foreign patients, then respect for the autonomy of the patients in the home country will tip the balance towards investing at home. So instead of donating money to Malawi to create better conditions to retain medical personnel, the UK should rather invest more in its own system so that it would not need to recruit Malawi medical personnel (Daniels, 2006).

We must try to proactively evaluate a certain evolution by collecting data on the projected effects. NaRanong and NaRanong have made a detailed estimation of the effects of the numbers of foreign patients on the demand for physicians in Thailand (NaRanong and NaRanong, 2011). Exact data are again important. The Apollo Hospital chain in India claims to have convinced 138 expatriate professionals to return to their country of origin by offering them more competitive salaries (Cortez, 2008). Hopkins et al. (2010) rightly point out that this number is so small that it will barely make a dent in the stream of physicians leaving each year from India. At the same time, Cohen claims that the effect will be limited in countries, like Thailand, where there are around a thousand private and public hospitals. The number of clinics for foreigners is relatively small and the medical staff working there makes up less than half a percent of the total medical capacity (Cohen, 2008). Finally, Pocock and Phua (2011) state that MT in Thailand does not pull doctors from rural areas but rather draws specialists from teaching hospitals in urban areas to the private clinics serving the foreign patients. Governments should monitor the evolution periodically in order to determine the effects on the different benchmarks and take measures to mitigate the undesirable consequences.

However, although only the destination of the medical personnel differs compared to the external brain drain, this difference may attenuate the negative effects. If no measures are taken to stop the external brain drain, the increase of incoming medical tourism may be a good thing for these countries. Clinics servicing foreign patients can function as a disincentive to expatriation. Again, the choice of reference point to determine the effects of medical tourism becomes important. Suppose, hypothetically, that 1000 doctors would leave the country if nothing changes compared to the present situation and suppose that when there is a steady flow of foreign patients coming into the country, only 250 doctors leave the country and 750 start working for clinics catering to foreign patients. Should we compare the outcome with the situation when nothing happens (medical personnel will flee to rich(er) countries), with the situation when clinics cater to foreign patients (medical personnel stay in the country and contribute at least to some extent to extra jobs in the clinics due to the health care activity) or with the situation when the government starts spending more money on the public health care sector so that wages can be raised and doctors are less likely to seek a better life elsewhere. In the first two situations, the public health sector is worse off because it loses medical capacity. Some people will be unhappy with both scenarios and will emphasise the need to do something about the brain drain in general. In other words, they compare with a situation in which all medical personnel stay and contribute to the public health care system. When compared with the ideal situation, incoming medical travel is obviously bad. However, when we use the first scenario as reference point, medical tourism is the least bad scenario and should be preferred. So, by the counterfactual test, the local population is harmed by MT (even when their condition has actually improved), but judged by the worsening test, they are not. The main challenge for the policy makers and ethicists is to find out which is the correct way to frame the problem.

Managing MT

An important problem in the evaluation is that MT is not seen as a health care system transformation. It is mainly treated as a means to increase export without really considering the effects on the public health care system. The general idea behind the promotion of MT is that foreign currency flows in and thus improves the general economic development, in other words, a 'trickle down' economics (Hopkins et al., 2010). Whether or not benefits from the MT sector spill over towards the total population depends on numerous factors. One of these will be whether the clinics are in foreign hands (mode 3 of the health services trade) or whether the profits go to the richest group. However, the main issue remains that these profits are not directed towards health care where the loss of fairness is felt most.

Most countries lack a coherent government policy to organise, regulate and control the process. 'While a carefully designed trade strategy in the health sector could have significant positive spill-over effects on the domestic supply and access to health services (in addition to the global positive impacts of trade), a poorly designed strategy could divert already scarce resources from developing countries' (Cattaneo, 2009). MT is almost exclusively seen as a source of income in the balance of payments. So how can MT be organised so that it maximises positive effects on health care provision in developing countries? Adequate accompanying measures have to be taken to prevent negative effects on the domestic health care system. Generally speaking, two options can improve the health care system: (1) the local patients benefit directly from the infrastructure and medical capacity constructed for the foreign patient and (2) part of the benefits generated by the trade is directed towards the public health sector. Countries like India, Singapore and Malta claim that they are able to offer a broader range of treatments and instruments (such as MRI scanners) to the indigenous patients because of the extra income generated through MT (Lunt et al., 2011). The second option can be implemented either at hospital level or at health care system level. The hospital could for instance use the income earned from foreign patients to cross-subsidise treatment of local patients. A similar idea was stipulated in the Public Trust Act in India: private hospitals had to spend up to 20 per cent of their resources on free health care. However, there is no control (Chinai and Goswani, 2005). Without active intervention by the government, it is highly unlikely that either of these options will materialise. The reason is relatively simple: this goes against the logic underlying the privatisation of health care. The new challenge for governments is to regulate the health sector at times when the participation of the private sector is constantly growing.

These policies and measures to redirect part of the profits of the health care trade should be structural and preferentially imposed by law or regulation. It cannot be left to individual hospitals to spend part of their benefits to subsidised care for the poor. This is unlikely to work in a highly competitive field since this will make such clinic more expensive and thus less attractive to foreign patients and investors. Highly mediatised 'charity' cases are even worse since they also violate justice principles by randomly selecting patients rather than following recognised rules of prioritisation. Some countries have laws but, as far as we know now, no one checks whether the clinics abide by the rules and no one forces them to apply the laws (Gupta, 2008). Nevertheless, in some cases, the system is adapted to generate direct extra income for the public sector. Some Australian public hospitals reserve some hospital beds for private foreign patients because they claim that one foreign patient generates enough extra income to treat three Australian patients (Cortez, 2008).

Thailand is a prime example on this point by a series of measures. It binds its medical graduates for a period of three years of compulsory public service, mostly in rural regions. It also gives considerable financial incentives to doctors in rural areas (Cattaneo, 2009). The important point demonstrated here is that active intervention and governmental regulation is necessary to prevent the situation from derailing.

Insurance companies

Medical ethics is a relatively young field but from its origins it has focused on the main players, that is, the patient and the doctor. Recently and reluctantly, ethicists (and in fact also doctors and patients) have come to realise that ignoring the context in which the patient–doctor dyad is situated is a major mistake. The third player that was included in the discussions was the government. The government, through its health care system and legal rules, started to shape and limit the options of both patients and doctors. However, in the meantime, things have become even more complicated: private insurance companies, corporate clinics (and chains of clinics), brokers and so on also increasingly intervene, leaving far less room for decision making by the original players (Turner, 2007).

It is obvious that funding will play a major role in the development of MT. Funding sources can be divided into two groups: the patient paying for the total cost out of pocket, and a third party, be it a public or private health insurer (Glinos et al., 2010). Within the last group, different mechanisms can be distinguished. Most publicly funded health systems apply to health services delivered and consumed on the territory of the country. Still, the recent directive on patient mobility in Europe changes this situation rather radically (European Parliament and Council, 2011). Patients, as a general rule, will be allowed to receive health care in another member state and be reimbursed up to the level of the costs that would have been assumed by the home country.

The directive wants to open up the market completely by abolishing the rule of prior authorisation for reimbursement. It is worth noting that the European Parliament and the Council in the conception of the framework for the application of patients' rights in cross-border health care do take into account possible negative effects of MT for the health care system. 'Overriding reasons of general interests' can be used to justify restrictions. If member states can provide evidence that the outflow of patients has a serious impact that undermines the planning of the local facilities, the rule of prior authorisation can be maintained. Similarly, if the inflow of foreign patients would lead to increasing waiting time for nationals, the destination state has the right to refuse treatment to foreign patients.

Private health insurers are commercial for-profit companies. Like all other companies, their activities are regulated by law and restricted by limits such as justice considerations and so on. They are, for instance, not allowed to discriminate on the basis of race or sexual orientation. The question then becomes to what extent these companies may offer (or coerce) people (to accept) specific arrangements. The reasoning may be that in private insurance, contractual clauses determine which health facilities the patients can use. If entering the initial agreement is voluntary, then why should there be any restrictions on the more specific elements? If people do not agree with the arrangement, they should not take the insurance. However, this is only correct when we are talking of health care interventions that are not part of what a society considers as reasonable health care.

The American Medical Association issued new guidelines on medical tourism (American Medical Association, 2009). These guidelines are very instructive as indications of the main concerns of the American doctors. The first guideline states: 'Medical care outside of the US must be voluntary'. This is indeed an important principle but what does voluntary mean? It could be argued that the incentives offered by the insurers come down to a 'coercive offer' or an offer one cannot refuse. At least three types of incentives are used: the insurer waives out-of-pocket expenses; premiums are lower when the costumer accepts to go abroad; and a percentage of the benefit that is made by going abroad is offered to the patient. At the same time, patients may face financial penalties such as high deductibles and co-payments if they do not accept to go abroad.

The price difference between health care facilities both within the country and outside the country will most likely lead to strong pressure from health insurers on clients to move to cheaper facilities (Mattoo and Rathindran, 2006). This pressure holds the following dangers: violation of the right to choose one's health care provider; loss of the right to opt for treatment at home and limitation of the type of treatment (cheaper, inappropriate interventions). At the same time, insurance companies may increase the patients' access to treatment, offer more guarantees for high-quality care for patients going abroad and make more information available for patients looking for treatment abroad.

As a final point of thought, we want to take a closer look at the price differences. Many articles quote the same tables that show that a coronary bypass costs ten times less in India than in the US. Since cost reduction is an important driver for MT, both for patients and insurance companies, it seems crucial to get a correct view of the situation. The most remarkable point is that all authors accept the data without questioning the assumptions. Recently, a study by Alleman et al. (2011) showed that the expected cost savings of overseas surgery were considerably lower than generally presented. The discrepancy can be explained by a difference in reference point. Alleman et al. used as a point of comparison the estimated cost of Medicare treatment instead of the price paid for private treatment. The assumption behind the use of data from private treatment is that most patients looking for cross-border treatment are uninsured (at least for the intervention they want or need) and thus would have to pay substantially more. Two points are important here: firstly, the assumption of the higher fees may not be correct (Melnick and Fonkych, 2008) and secondly, the total costs for overseas treatment (including fare, hotel and so on) may be comparable with Medicare costs in the US. Both elements would have serious implications for the estimated growth of MT to developing countries.

Conclusion

Privatisation and globalisation of health care are relatively new phenomena of which the impact on existing health care systems, on the status of health care, on views of patients and responsibilities of doctors is unknown. However, it is clear that the effects may be fundamental. This should urge us to exercise caution and closely follow emerging trends. The complexity of the development of medical travelling as a health care reform should also encourage a critical attitude that steers clear from sweeping statements that generalise local findings to universal laws. The judgement is still out as far as the final consequences are concerned: the development of medical travelling may be beneficial overall depending on the context. Still, there are many intriguing ethical issues to be resolved: beneficial for whom? Compared to which situation? According to what standard? These matters will not be decided by empirical evidence. A stronger and more developed theory of justice is needed to answer these questions.

References

Alleman, W. B., Luger, T., Reisinger, S. H., Martin, R., Horowitz, D. M. and Cram, P. (2011) 'Medical tourism services available to residents of the United States' *Journal of General Internal Medicine* 26, 5, 492–7.

American Medical Association (2009) 'New AMA guidelines on medical tourism' http://www.ama-assn.org/ama1/pub/upload/mm/31/medicaltourism.pdf date accessed 19 September 2012.

Cattaneo, O. (2009) *Trade in Health Services. What's in it for Developing Countries?* (The World Bank: International Trade Department).

Chinai, R. and Goswani, R. (2005) 'Are we ready for medical tourism?' *The Hindu*, 17 April 2005 http://www.thehindu.com/thehindu/mag/2005/04/17/stories/2005041700060100.htm date accessed 15 November 2011.

Cohen, E. (2008) 'Medical tourism in Thailand' *AU-GSB e-Journal* 1, 1, 24–37.

Cohen, G. I. (2011) 'Medical tourism, access to health care, and global justice' *Virginia Journal of International Law* 52, 1, 1–56.

Commonwealth (2003) *Commonwealth Code of Practice for the International Recruitment of Health Workers*, http://www.thecommonwealth.org/shared_asp_files/

uploadedfiles/%7B7BDD970B-53AE-441D-81DB-1B64C37E992A%7D_CommonwealthCodeofPractice.pdf date accessed 23 November 2011.

Cortez, N. (2008) 'Patients without borders: The emerging global market for patients and the evolution of modern health care' *Indiana Law Journal* 83, 71–132.

Daniels, N. (1985) *Just Health Care* (Cambridge: Cambridge University Press).

Daniels, N. (2006) 'Equity and population health: Toward a broader bioethics agenda' *Hastings Center Report* 36, 4, 22–35.

Daniels, N., Bryant, J., Castano, A. R., Dantes, G. O., Khan, S. K. and Pannarunothai, S. (2000) 'Benchmarks of fairness for health care reform: A policy tool for developing countries' *Bulletin of the World Health Organisation* 78, 6, 740–50.

European Parliament and the Council of the European Union (2011) 'Directive 2011/24/EU of the European Parliament and of the Council of 9 March 2011 on the application of patients' rights in cross-border healthcare' *Official Journal of the European Union* 4 April 2011, L. 88/45-L. 88/65.

Feinberg, J. (1986) 'Wrongful life and the counterfactual element in harming' *Social Philosophy and Policy* 4, 1, 145–78.

Glinos, A. I., Baeten, R. and Maase, H. (2010) 'Purchasing health services abroad: Practices of cross-border contracting and patient mobility in six European countries' *Health Policy* 95, 103–12.

Gupta, S. A. (2008) 'Medical tourism in India: Winners and losers' *Indian Journal of Medical Ethics* 5, 1, 4–5.

Healy, C. (2009) 'Surgical tourism and the globalization of healthcare' *Irish Journal of Medical Sciences* 178, 125–7.

Hermans, H. (2008) 'Cross-border health care in the European Union: Recent legal implications of "Decker and Kohll"' *Journal of Evaluation in Clinical Practice* 6, 431–9.

Hopkins, L., Labonte, R., Runnels, V. and Packer, C. (2010) 'Medical tourism today: What is the state of existing knowledge?' *Journal of Public Health Policy* 31, 185–98.

Labonte, R. (2004) 'Globalization, health and the free trade regime: Assessing the links' *Perspectives on Global Development and Technology* 3, 1/2, 47–72.

Lunt, N., Smith, R., Exworthy, M., Green, T. S., Horsfall, D. and Mannion, R. (2011) *Medical Tourism: Treatments, Markets and Health System Implications: A Scoping Review* (Paris: Directorate for Employment, Labour and Social Affairs, OECD).

Mattoo, A. and Rathindran, R. (2006) 'How health insurance inhibits trade in health care' *Health Affairs* 25, 2, 358–68.

Mechanic, D. (2002) 'Disadvantage, inequality, and social policy' *Health Affairs* 21, 2, 48–59.

Meghani, Z. (2011) 'A robust, particularistic ethical assessment of medical tourism' *Developing World Bioethics* 11, 1, 16–29.

Melnick, A. G. and Fonkych, K. (2008) 'Hospital pricing and the uninsured: Do the uninsured pay higher prices?' *Health Affairs* 27, w116–22.

Morreim, H. E. (1983) 'Conception and the concept of harm' *Journal of Medical Philosophy* 8, 137–57.

Ormond, M. (2011) 'Shifting subjects of health-care: Placing "medical tourism" in the context of Malaysian domestic health-care reform' *Asia Pacific Viewpoint* 52, 3, 247–59.

Pennings, G. (2007) 'Ethics without boundaries: Medical tourism' in R. Ashcroft, A. Dawson, H. Draper and J. McMillan (eds) *Principles of health care ethics*, 2nd edn (London: John Wiley and Sons).

Pocock, S. N. and Phua, H. K. (2011) 'Medical tourism and policy implications for health systems: A conceptual framework for a comparative study of Thailand, Singapore and Malaysia' *Globalization and Health* 7, 12.

Pogge, T. (2005) 'Severe poverty as a violation of negative duties' *Ethics and International Affairs* 19, 1, 55–83.

NaRanong, A. and NaRanong, V. (2011) 'The effects of medical tourism: Thailand's experience' *Bulletin of the World Health Organisation* 89, 336–44.

Smith, D. R., Chanda, R. and Tangcharoensathien, V. (2009) 'Trade in health-related services' *The Lancet* 373, 9663, 593–601.

Turner, L. (2007) '"First world health care at third world prices": Globalization, bioethics and medical tourism' *BioSocieties* 2, 303–25.

Turner, L. (2010) '"Medical tourism" and the global marketplace in health services: US patients, international hospitals, and the search for affordable health care' *International Journal of Health Services* 40, 3, 443–67.

UNCTAD Secretariat (1998) 'International trade in health services: Difficulties and opportunities for developing countries' in UNCTAD-WHO (eds) *International Trade in Health Services: A Developmental Perspective* (Geneva: UNCTAD) http://www.ictsd.org/issarea/services/resou.rces/Docs/WHO-UNCTAD1.pdf date accessed 8 December 2010.

World Health Organisation (2010) *WHO Global Code of Practice on the International Recruitment of Health Personnel* http://apps.who.int/gb/ebwha/pdf_files/WHA63/A63_R16-en.pdf date accessed 13 January 2011.

Youngman, I. (2009) 'How many American medical tourists are there?' *International Medical Travel Journal* http://www.imtj.com/articles/2009/how-many-americans-go-abroad-for-treatment-30016/?locale=en date accessed 12 June 2010.

12
Habermas, Transnational Health Care and Cross-Culturalism

Tomas Mainil, Vincent Platenkamp and Herman Meulemans

The key arguments of this chapter are as follows:

- Transnational health care (THC) is a futuristic, coordinated and professionalised provision of cross-border health care and medical tourism services. It is an emergent field (Lunt et al., 2011) full of many opportunities, but it also holds risks.
- A conceptual basis is lacking, therefore the application of an established thought model – Habermas' action theory – would be beneficial for understanding the nature and dynamics of THC.
- The purpose of this chapter is to introduce the legacy of Jürgen Habermas and adapt it to the context of THC, in order to show the complex cross-cultural dynamics that play a role in THC.

Introduction

In this chapter we introduce the basics of Habermas' general action theory with its communicative and strategic actions/life-world and system. Then we link his theory to the dynamics present in transnational health care (THC) (Mainil et al., 2012ith its market/health consumer/ethical/professional perspectives. Furthermore, we focus on a case study in THC, showing at a micro-level how there are tensions between communicative and strategic actions. The role of cross-cultural management is introduced to solve these tensions. Finally, the relationship between cultural management, THC and Habermas' framework is discussed.

Basic concepts of Habermas' theoretical framework

In the globalising world of THC, new areas of intercultural communication emerge that lead to particular directions in the discourses about THC. To understand these new directions employing the theoretical framework of the German philosopher Jürgen Habermas (Habermas, 1982) seems to be of relevance. Habermas tried to find an answer to the question of why so many things have gone wrong with the process of rationalisation that has accompanied the emergence of modern Western societies. Since we entered the era of globalisation, this question seems to have become relevant for the whole, interconnected globe, in many situations of everyday life, with a tension between the global and the local. A characteristic of the Western situation Habermas reacted to is the modernisation of the whole of society. A new instrumental rationality that made many things in life more efficient and effective, replaced the old coordination mechanisms of, for example, religious resources in pre-modern times. On the one hand, this new situation provides more welfare based on this type of instrumental rationality. On the other hand, it goes along with what Habermas calls various 'pathologies' in late-capitalist societies, such as alienation, fragmented identities, materialism, consumerism, the replacement of citizens by consumers and the impressive influence of a massive culture industry. In order to analyse the ambivalence in modern society, Habermas distinguished two types of actions in his general action theory, which enables a strong characterisation of the situation that can be extended to the global–local nexus in THC.

First, there is *communicative action*. One of the basic questions of sociology has always been 'how is social order possible?' Through communicative action, one can answer by reference to the common definitions of the situation or reality that are made in everyday communication. These common definitions not only refer to the objective situation of the actors but also to the norms that are obligatory to them and to the truthfulness of their utterances. In their daily communication, people come to agreements by bringing frames of interpretation into discussion that are criticisable. Through argumentation that is free of power and oppression, agreements will emerge, according to Habermas, with this type of communicative action. The *life-world*, another crucial concept of Habermas, constitutes the horizon against which these communicative processes take place. It consists of non-problematic, mostly diffuse, background convictions that are the source of the attainment of common definitions of situations of communicative actions. Life-worlds contain a stock of experiences of previous generations in the form of interpretations of reality, constraining norms and fixed interpretations of needs. From within these life-worlds an argumentative type of discourse takes place that gives meaning to the everyday life of its participants, and constitutes a coordination mechanism for their actions because of their shared reality by common definition.

This, of course, is an ideal situation that is simultaneously challenged by a second type of action, *strategic action*. Everyone knows that he or she will sometimes be forced to accept a definition of the situation. This can be the consequence of strategic action, which is not oriented towards shared understanding but to results. Now, on the one hand this strategic action has brought about, in particular for Western society, a great deal of progress, strong political power and a flourishing economy. The reason is easy to understand: based on efficiency and effectiveness strategic action creates a much stronger coordinated type of action that speeds up and replaces the argumentative type of action by communication. The type of instrumental rationality that results from it has created a money-based, interdependent world economy, as we have learned to understand it through the concept of a global village. The same goes for the power mechanisms that go with this type of rationality and that have created our global political system, with its divisions worldwide. This economic and political *system* is another crucial concept in Habermas' theory, introduced in opposition to his life-world. Although many good things originate from the efficient and effective coordination of this system, it also has caused a crucial tension with the much more vulnerable, argumentative, processes of the life-world.

Colonisation and fragmentation of the life-world in-between the global and the local

Life-worlds can only be maintained by communicative action. When this is not functioning well, as in modern society, we witness the colonisation and fragmentation of life-worlds. Maintenance by communicative action should result in common frames of interpretation, solidary, group formation and responsible individuals. Ideally speaking, the identity of a society would become, then, completely dependent on argumentatively steered processes of interpretation by its actors. There is a development in this direction, according to Habermas. But there are disturbances of these processes that can be very serious, because of the destructive influences they have on the processes of interpretation. On the one hand, life-world processes enable systems that are based on efficiency and effectiveness. And as long as they stay within the control of these interpretations, they are the basis of progress in our modern society. At the same time, however, this systemic influence becomes independent and starts to replace the interpretive coordination in modern life-worlds. States and markets deliver other dominant coordination mechanisms in the life-worlds of people, and cause a fragmentation and colonisation of these life-worlds through the growing importance of money and power over communicative interpretation. Markets and bureaucracy start to dictate the coordination of life-worlds to an unacceptable degree. For example, in higher education the knowledge economy often seems to have become the defining factor of its success. International rankings are followed, and the discussion about, amongst others, what education should imply for responsible world citizens seems to be replaced by them. Managers decide the costs and benefits of education and research, and are less interested in the academic mentality of doing careful research for the long term. This is unacceptable, because in their life-worlds humans should remain in control through their own frames of interpretation of what their world of higher education looks like or should look like, instead of being outvoted by the market mechanisms of the day. When this happens in a systematic way, life-worlds will be colonised by this systemic influence of money and power and their frames of interpretation will become fragmented and will lose their meaning and power for coordination.

This type of phenomenon has been analysed and extended by many theorists who follow Habermas in his explanation of the modernisation of the Western world. In this study his basic concepts, as explained earlier, will be applied to the situation of THC on a global scale, far beyond the Western world. In a globalising world this seems to be a logical next step.

Making a bridge between the rationale of THC and Habermas' framework

In a globalising THC, the intermediary organisations in-between the system and life-world seem to form the place where the tension between communicative and strategic actions is mostly felt. For Bourdieu (1984), the stakeholders that are committed to this relatively new field constitute the 'new petite bourgeoisie' that emerges in a post-modern market situation (Featherstone, 1991). They are situated in-between production and consumption, and symbolic production is central to their activities, which frequently means the use of advertising imagery, marketing and promotional techniques. Translated to Habermas' concepts, this would refer to an intensified struggle between the system and life-world, in which the human values from within their life-worlds are coming under the pressure of the demanding wishes of highly efficient and effective markets. One of the given facts in the medical tourism market is this drive to be highly efficient, leaving the gate open for severe strategic action. Therefore, let us look at the perspectives present in THC (adapted from Mainil et al., 2011) and link them to Habermas' action theory.

One can detect several perspectives present in the system of THC.

1. *Ethical perspective (demand driven)*: this relates to the specific character of the relationship between caregiver and patient, which is different from other service-related relationships between providers and customers.
2. *Market perspective (supply driven)*: this relates to the industry of medical tourism that is based on making profit and the creation of valuable business models.
3. *Health consumer/patient perspective (demand driven)*: this relates to the specific character of the international patient as both a citizen with rights and a consumer with wants and needs.

4. *Professional perspective (supply driven)*: this refers to the specific role of the professional in the multi-actor field of medical tourism and cross-border health care.

First of all there is an ethical perspective present in THC. Stakeholders are offering (medical) services but do have to take into account the *ethical perspective* of treating a patient (demand-side perspective), which is different in nature compared with selling a hotel package to a prospective tourist. Although this is clear, the existence of brokers in the field, such as medical tourism facilitators and traders in the organ market, does not always allow this means of ethical reasoning, because the *market perspective* has an established tradition in the developing sector of medical tourism. Businesses and hospitals see ways of making profit on the basis of prospective streams of international patients, and translate these into marketing and business plans (supply-side perspective). The medical tourism sector is built on the generation of magical numbers of international patients: no real quantifiable data exist. The websites of the brokers show content that is not always built on truthful medical quality (Lunt et al., 2010; Mainil et al., 2010; Crooks et al., 2011; Lunt and Carrera, 2011; Turner, 2011a). There is a perseverance to develop an international health care market: a market perspective that is known as a trade in health services (Smith et al., 2009). International patients are seen by the sector as consumers of health services. It is known that the development of health literacy skills leads to more empowerment (Nutbeam, 2000). This could result in the development of pro-active searching skills on the Internet, searching for health information, along with developing another relationship with medical professionals. This image is partially true, because in many cases patients do not have much choice in working out what health options they have. However, on the assumption that a THC sector will arise that is transparent, professional and based on a skilled network, health consumers (Iriart et al., 2011) will get the chance to make appropriate choices and decisions: a health *consumer/patient perspective*. Consumerism is seen here as having almost the same connotation as citizenship, where patient rights do have a large part to play in the empowerment of patients (demand-side perspective). Finally a specific *professional perspective* is also developing. Professional elements from different sectors are combined and intertwined in THC: the tourism industry and the medical sector. One cannot deny that a certain touristification takes place in the medical sector by means of medical tourism. How can the professional medic stay responsive in a sector ruled by packages, Internet marketing and a striving for commercial accreditation (supply-side perspective)?

In order to make these four perspectives, prevalent and deterministic in THC, of use within the framework of Habermas, one needs to position them as strategic action and/or communicative action. Therefore we want to form the thesis that being ethical and being a knowledgeable consumer/citizen as a patient are forms of communicative action. On the other hand, marketing and professionalising the services would be seen as forms of strategic action. One could argue that it is possible to see a sustainable consumerism as a form of communicative action, without breaking the rules of Habermas' view on the development of consumerism. It is the life-world in which the patient operates, based on a set of communicative rules: his or her patient rights. When an international patient seeks information as a responsible user of health services abroad, he or she hopes that the information received is based on common understandings regarding how to communicate truthfully between the consumer and a medical supplier within THC. Strategic action, in this case, would be for the medical supplier to market their services in such a way that their organisation benefits from it, but, in order to do so, the supplier abandons the equal bond between doctor and patient and lets commercialisation of health services rule. Systemic occurrences are instigating a medical tourism market, but one based on strategic actions; the market is creating itself with a view to monetary and business objectives. Also, what one considers as ethical in health matters should be seen as a joint exercise in communication within a health community, with dialogical rules of engagement. As a counterpoint, the way an organisation professionalises itself could also be determined fully by strategic action. Again this is based on non-democratic processes that are efficient in making profit but could be dysfunctional for the patient as a global consumer. Apart from this, but not of relevance in the context of this chapter, it excludes the perspectives of the patients and other stakeholders in the local health care systems around the globe.

The case of the international office in the UKE Hamburg

The intent now is to show a combination of Habermas' action theory and THC perspectives in a single case study of a hospital. The Universitätsklinikum Hamburg Eppendorf hospital (UKE) has been attracting international patients for more than a decade. Its service-oriented style suggested that an approach on behalf of the patient was necessary. Therefore a department of non-medical services was created: the International Office (IO). Semi-structured face-to-face interviews were held with the director, marketing manager, case manager and interpreters of the IO in order to investigate how such a department operates. On average, interviews lasted one hour. Interviews were recorded and analysed by means of thematic analysis.

As an intermediary between the patient and the medical staff, the IO is seen as operating between producers and consumers of medical services, from a service-oriented perspective. One can observe a combination of strategic and communicative actions: on the one hand, the strategic perspective of the functionalities of a hospital with regard to professionalism and, on the other hand, the mutual understanding between international patients and the members of the IO:

> we consider ourselves as a service point or link between the patient and the clinic ... we function as a link between patient and the departments, we often speak to both of them, we always try to do the best we can in the interest of the patient, but also try to do our best in the interest of the clinic. That is not always easy.
>
> (Management IO)

As an organisational unit within the hospital, its members are mediators and have to show instances of agency on behalf of the international patient, while at the same time they have to take into account the strategic actions of the medical staff and departments. But by working with the patient as an equal partner, the IO installs communicative actions, while also making efforts to include professional codes of practice. For example, the IO tries to help patients by reducing waiting times as this is sometimes expected by international patients, although under the professional code of practice the hospital needs to treat every patient, German and international, equally (Management IO).

With regard to this mediating function, there are several professional roles present in the UKE department. They serve as an example of how other professions can contribute to the guidance of a patient. First of all there are interpreters, who have a critical role in the process. They are the human face of the hospital, and the liaison between the international patient and every service that is provided. Mutual cross-cultural understanding between patient and interpreter inserts a gap in the system of the hospital; a life-world is created in the view of the patient, a consumer with citizen rights:

> The role of the IO is to function as a link between the patient and treatment facility/doctor. Because of the special circumstances the patient is in and the sometimes delicate topics that have to be translated, often a bond is developed between the patient and the translator. The patient mostly fully trusts the translator and sees him as a true friend abroad. Sometimes even after the patient has already got back home, letters are written or visits are scheduled. This is then of a totally private nature and has nothing to do with the professional relationship when they started to know each other.
>
> (Interpreter 1)

This close relationship with the patient sometimes results in a non-professional relationship that of course is related to the specific character of the international patient; frictions between strategic and communicative actions are inherent to the professional role of the interpreter.

With the medical facilitator as an agent of commercialised health-related services (Snyder et al., 2011; Turner, 2011b), the UKE department at some point shows its capacity to resolve disputes. It is critical in using the services of the facilitators, who sometimes try to indulge their own agency by interfering in the process between interpreter and patient:

> the relationship is sometimes difficult, because some facilitators try to get involved in the process too much, they try to change things in the clinics, they speak to doctors directly, which we don't like at all. They sometimes function as interpreters, which we really don't like at all; we have medical translators, they know what they are doing, they know the medical terminology, while the facilitators have little knowledge of medicine. So we don't like them to interact too much, which sometimes they do.
>
> (Management IO)

Next to the interpreters, the role of the case manager is decisive for the mediating role between hospital and patient. The case managers are the regulators of the system. They are there from the early stages of contact to ending the services. Additional services are less common in a health care setting: departures from selected hotels, contracts with airlines, special oriental food needs and more specialised tourism possibilities. This indicates the relevance of other market-driven services for a sector such as THC.

> if the patient asks us to do that, we do it, we have a website which shows the hotels which we have contracts with. We also have a contract with Lufthansa for special rates. Usually patients do this by themselves, or their facilitators do it for them; if they ask us to do it, we do it. We have some external partners who we work with ... Mainly travel and lodging, sometimes people want to do a tour around the town; we can organise that for them, or tell them where they can get it easily, or they want to rent a car, or something special, where they can go shopping ... Yes there is, but we offer that here, we have special Arabic food for our patients that we offer, we can provide Kosher food, not done at the hospital but we can order that, if they ask for that it is not a problem; we also have special foods for the Russian patients, they sometimes like special Russian food, so we can offer that.
>
> (interview Management IO)

Sometimes the opposite occurs: patients are consumers in the strategic sense, forcing the hospital and the IO to focus on efficiency, by speaking about their shopping behaviour to international hospitals and the experiences they have had:

> Yes, most of the time they tell us where they have already been, they compare us with other hospitals in Germany or in other countries and tell

> us immediately what they like about this hospital. They try to describe it as if they heard that it was better somewhere else. They try to compare anyway.
>
> (Interpreter 2)

Patients also show themselves as owners of their own health by asking for information that is sometimes difficult to get. Again this seems to be a collision of market (strategic) and consumer (communicative) forces: in order to be efficient as an organisation, clarity of the risks needs to be provided to create a field of trust with the patient:

> Patients ask for success rates of treatments, but is very difficult, because it is based on the health reports of patients, which is not the same as the current status of the patient, which is not known to us.
>
> (Case Manager)

The agency character of the department is enhanced by its capacity to offer non-medical services, which do play a role in its mediation between patient and hospital. The department is confronted with the otherness of some of its patients, therefore the agency also includes guarding the cross-cultural identity of its hospital while maintaining relations with other patients and medical staff:

> Arabic patients in particular do bring a lot of people with them, they sometimes have three, four or five people accompanying them, and those people like to hang out with the patient, they think it is their duty to stay with the patient all the time, and they can be very loud and this is challenging for the staff and other patients who are close to those patients. And we try to mediate – and this is done by the interpreters – who then explain that they should be a little more quiet or go inside, because all the patients need to have a good stay in the hospital.
>
> (interview Management IO)

By interviewing the management, it became obvious that a different ethical standpoint is needed for THC, especially in terms of marketing the services of the department and its hospital:

> Yes, you need to think about the morals and the ethics: this always needs to be considered, in all the marketing efforts we do; what we do not do is offer something we cannot hold to, what I said about the discrepancy between the ideal world of marketing and practice, we are not saying 'when you come here everything will be perfect and you will be healed'... you need to be very careful, because you are dealing with very ill people and you always have to consider that.
>
> (interview Management IO)

In conclusion, we can state that a service department in a medical environment does need a different dynamic than other commodified service areas. Different ethical standpoints are needed, inherent to the community action-based medical context. A market perspective is possible, but it is about mediating between two or more other agency groups. Being a consumer as an international patient incorporates larger risk and effort than being a consumer as a local patient. Finally the professional roles are specifically linked to these other agency groups, such as the medical staff, medical facilitators and international patients, and suffer from a larger complexity and cross-cultural capacity than present in other sectors. All the narratives from the IO suggest, at this micro-level, that the unit needs to balance patients' rights and a more commercial strategic action model. This balance also needs to be maintained when looking at the patients as consumers: every patient needs to be treated with equity, but because of the large effort needed to treat these patients, a combination of strategic and communicative frames is present. This is intertwined with the professional role of the IO: designed to service international patients and to be efficient and strategic, but nested within a European concept of how to secure the rights of patients and their citizenship. The market perspective requires the organisation to move into and to act in a commodified business arena, but at the same time it is part of a university hospital with a rich tradition based on a communicative, historic life-world.

By applying a form of cultural sensitivity and adaptation (Liu et al., 2011), internalised by the staff of the IO, this could be a tool for the THC sector to overcome the gap between strategic action and the life-worlds of communicative action. Being culturally aware of the other (Said, 1979; Hall and du Gay, 1996), in this case the international patient can align the patient as a citizen with the more strategic targets of the organisations present in THC. This alignment would be the acceptance of the differences in cultural norms and thought frames between patient and professional in an international context. To accept that an Arabian patient is different in a German cultural context, and to adapt your supply of services according to this cultural awareness, is a way of creating a new globalised life-world on the basis of developing an intercultural identity (Liu et al., 2011): it aligns your efficiency as an organisation towards the rules of a clash of life-worlds, a culturally defined concept of health-care provision.

Discussion

As an important actor in the THC system, we observed an international patient department that works in direct proximity with a reputable hospital. One can see that the department has developed its own professional profile and means of communicating and dealing with international patients, in-between these patients and the medical professionals. Being very close to the medical provision of the medical services, it has to balance market-driven perspectives and higher standards of ethics and professionalism. This makes it unique but shows a case for renewal within the THC sector, close to the medical paradigm.

In the IO, a combination of strategic and communicative actions occurs: on the one hand, the strategic perspective of the functionalities of a hospital and its professionalism; on the other hand, a mutual understanding between international patients and the members of the IO. In its mediating function, it has to fulfil several professional roles, in which the tension between communicative and strategic actions becomes obvious. As interpreters with a critical role, it provides the human face of the hospital and the liaison between all hospital services and the international patient. As owners of their health, patients also want the organisation to create a field of trust with them. The patient trusts the interpreter in his or her emergent life-world as a patient with citizen rights. It may become a close relationship that extends beyond non-professional aspects, which can lead to tensions between strategic and communicative actions within the professional role of the interpreter in the IO. A medical facilitator in close contact with the patient can demonstrate some capacity for dispute. Sometimes he or she may speak to doctors directly and interfere with medical actions that he or she has little knowledge of. Other more market-driven services, additional to the health care setting, show their relevance for THC as well, where the role as case manager between hospital and patient is concerned. Non-medical services include confrontation with the otherness of its international (for example, Arabic) patients compared with other patients, which points to the cross-cultural identity of the hospital and the relationships with other patients and the medical staff.

The professional roles mentioned are all specifically linked to the other agency groups involved, such as the medical staff, medical facilitators and international patients. Their professional character suffers from a larger complexity and cross-cultural capacity than present in other sectors. From a cultural perspective, our world is a hybrid world (Appadurai, 1996; Hollinshead, 1998), with many interacting life-worlds all over the globe. The situation within THC is not different in this respect and the international patient department has to handle this complexity. Different agency groups create a new dialogue with many cross-cultural constraints. In this field of communicative action some basic cross-cultural skills could be of great help to facilitate the necessary coordination for this type of action. In their different life-worlds, participants make use of different cultural and other sources. An example of such skills, therefore, could be the empathic capacity to change perspectives, which means that one should listen carefully and try to take the role of the other while listening. This capacity can be trained to a certain degree and its cross-cultural application would provide much more serious coordination in the field of communicative action. Other concepts of cross-cultural management, such as changing perspectives, contextual approaches, self-reflexivity and acknowledging the patient as an equal partner, could help in facilitating this coordination, which seems to be so crucial in this world of hybrid complexity. Particularly in the much debated area of ethical discussions, norms and values may originate from various religious and culturally related sources. Here too, there is still too little attention to the cross-cultural aspects of this type of normative discourse in THC. This is true in spite of the quickly developing influence of strategic forms of action that take place at the same time. In this chapter we have also looked at the expanding market discourse in the field, based on efficiency and effectiveness as criteria of coordination. Mainil et al. (2011) refer to a breakdown in global discourse on THC in this respect. After an initial stage of domination by an ethical type of discourse, there was a sort of breakdown in this global discourse towards a market discourse. Such a breakdown always seems to suggest the emergence of more strategic actions in the field of THC. The market-driven constraints become stronger in the field of health care, which will always remain sensitive to a certain degree to that other type of (communicative) action. In this new development there is a hidden threat of replacement of a communicative coordination mechanism, with its complex cross-cultural constraints, by a seemingly more efficient and effective coordination mechanism that originates in strategic action. This does not mean that, where strategic action is needed, it should not be accepted. It means that where communicative action is needed it should not be colonised by strategic action, as meant by Habermas' concepts.

Conclusive remarks

The IO, as an intermediary, shapes the cross-cultural relationship between international patients and the hospital. Therefore, it looks as if two scenarios for professionalism are occurring in the context of THC. Firstly there is a scenario with a sustainable professional context-controlled model, where market, quality and sustainability meet each other, offering intermediaries that have a quality-oriented mediating role. In this scenario there is a sophisticated balance between communicative and strategic actions. Secondly, there is a model where there is only a market-driven sector, close to tourism, where each of the intermediaries has its own market drivers, with unrestricted production of uncontrolled mediating sources. According to this model, the life-worlds of the participants of THC will be colonised, with the consequences referred to in this chapter.

Acknowledgements

We would like to express our gratitude to the professionals of the International Office of the UKE Hamburg Eppendorf University Hospital.

References

Appadurai, A. (1996) *Modernity at Large: Cultural Dimensions of Globalization* (Minneapolis: University of Minnesota Press).

Bourdieu, P. (1984) *Distinction: A Social Critique of the Judgement of Taste*, translated by R. Nice (Cambridge, MA: Harvard University Press).

Crooks, V. A., Turner, L., Snyder, J., Johnston, R. and Kingsbury, P. (2011) 'Promoting medical tourism to India: Messages, images, and the marketing of international patient travel' *Social Science and Medicine* 72, 726–32.

Featherstone, M. (1991) *Consumer Culture and Postmodernism* (London: Sage).

Habermas, J. (1982) *Theorie des kommunikativen Handelns* (Frankfurt am Main: Suhrkampf).

Hall, S. and du Gay, P. (eds) (1996) *Questions of Cultural Identity* (London: Sage).

Hollinshead, K. (1998) 'Tourism, hybridity and ambiguity: The relevance of Bhabha's third space cultures' *Journal of Leisure Research* 30, 1, 121–56.

Iriart, C., Franco, T. and Merhy, E. E. (2011) 'The creation of the health consumer: Challenges on health sector regulation after managed care era' *Globalization and Health* 7, 2, www.globalizationandhealth.com date accessed 7 February 2012.

Liu, S., Volcic, Z. and Gallois, C. (2011) *Introducing Intercultural Communication, Global Cultures and Contexts* (London: Sage).

Lunt, N., Hardey, M. and Mannion, R. (2010) 'Nip, tuck and click: Medical tourism and the emergence of web-based health information' *The Open Medical Informatics Journal* 4, 1–11.

Lunt, N. and Carrera, P. (2011) 'Advice for Prospective Medical Tourists: Systematic review of consumer sites' *Tourism Review* 66, 57–67.

Lunt, N., Smith, R., Exworthy, M., Green, S. T., Horsfall, D. and Mannion, R. (2011) *Medical Tourism, Treatments, Markets and Health System Implications: A Scoping Review* (OECD: Directorate for Employment, Labour and Social Affairs, available online).

Mainil, T., Platenkamp, V. and Meulemans, H. (2010) 'Diving into contexts of in-between worlds: World making in medical tourism' *Tourism Analysis* 15, 743–54.

Mainil, T., Platenkamp, V. and Meulemans, H. (2011) 'The evolving discourse of medical tourism in the media: In search of the rupture' *Tourism Review* 66, 31–44.

Mainil, T., Van Loon, F., Dinnie, K., Botterill, D., Platenkamp, V. and Meulemans, H. (2012) 'Transnational health care: A quest for a global terminology' *Health Policy* 108, 1, 37–44.

Nutbeam, D. (2000) 'Health literacy as a public health goal: A challenge for contemporary health education and communication strategies into the 21th century' *Health Promotion International* 15, 3, 259–67.

Said, E. (1979) *Orientalism: Western conceptions of the Orient* (London: Penguin Books).

Smith, R. D., Rupa, C. and Viroj, T. (2009) 'Trade in health-related services' *The Lancet* 373, 593–601.

Snyder, J., Crooks, V. A., Adams, K., Kingsbury, P. and Johnston, R. (2011) 'The "patient's physician one-step removed": The evolving roles of medical tourism facilitators' *Global Medical Ethics* 37, 530–4.

Turner, L. G. (2011a) 'Quality in health care and globalization of health services: Accreditation and regulatory oversight of medical tourism companies' *International Journal for Quality in Health Care* 23, 1, 1–7.
Turner, L. (2011b) 'Canadian medical tourism companies that have exited the marketplace: Content analysis of websites used to market transnational medical travel' *Global Health* 14 October, 71, 40.

13

The Impact of Medical Tourism in Low- and Middle-Income Countries

Melisa Martínez Álvarez, Richard D. Smith and Rupa Chanda

The key arguments of this chapter are as follows:

- Medical tourism flows have reversed in recent years with increasing numbers of patients travelling from high- to low- and middle-income countries; drawn by cheaper prices, greater availability and increased privacy.
- It is difficult to know precisely how much medical tourism takes place because data are not systematically collected. However, the literature suggests that it is happening on a large, and increasing, scale.
- Low- and middle-income countries providing medical tourism companies marketing cross-border medical travelservices may benefit from generating foreign exchange, attracting – and retaining – health professionals and improving facilities and quality of care.
- These countries also risk diverting resources to cater for foreign patients that can bring in higher revenues, thereby neglecting the local population.
- There are three types of trade agreements countries can engage in when providing medical services to international patients: multi-lateral, regional and bi-lateral.
- Bi-lateral trade offers countries the greatest scope to capitalise on the benefits and reduce the risks of engaging in medical tourism, as seen by a case study from a potential UK–India relationship.
- More data are needed on the scale of medical tourism and its effects, to increase the evidence available for policy makers and to allow for decisions to be more evidence-based.

Introduction

Medical tourism, of the scale and scope currently witnessed, is a product of contemporary globalisation. Globalisation is the process by which economies, societies and cultures become inter-connected in a global network. Economic factors, such as foreign direct investment, migration and trade are tautologically linked with globalisation. In the case of health systems, it is international trade, and the increased liberalisation – or opening – of markets, which has led to the practice of medical tourism. It is important then to understand the trade relationships underpinning developments, and the economic as well as health and health system implications, as it is economic factors that largely drive medical tourism.

This book has thus far dealt with different types of medical tourism, the types of patients that engage in this practice and the ethical and cultural considerations that need to be taken into account when studying medical tourism. This chapter will explore the effect medical tourism has on the low- and middle-income countries that provide these services. Many low- and middle-income countries have opened up their health systems to international trade, mainly in the form of outsourcing and treating foreign patients, lured by the prospects of generating revenue and increasing employment. However, as we shall see in this chapter, opening health systems to medical tourism may also have some negative unintended consequences for these countries.

It should be noted that in this chapter we consider low- and middle-income countries to be predominantly nations that export health services to foreign patients (that is, countries which attract medical tourists) from high-income countries. Historically there has been a small level of trade in the opposite direction (the classic case of wealthy individuals from poor countries travelling to Harley Street in London or the Mayo Clinic in the US for care), but this is beyond the scope of this chapter.

The chapter will provide an overview of global flows of medical tourists and will critically review issues associated with the quantity and quality of the data available. This will be followed by an account of the impact medical tourism can have on low- and middle-income countries, both the benefits and the risks countries face when offering health services to foreign patients. This will be followed by a discussion of the different types of trade agreements under which medical tourism can happen, and how these influence the impact this practice has on low- and middle-income countries. The chapter will conclude with a case study illustrating how issues arising from medical tourism can be tackled through a bi-lateral trade agreement, focusing on the UK and India.

Do medical tourists travel to low- and middle-income countries?

Medical tourism is far from being a recent phenomenon; people have been travelling to access health care in faraway places for centuries. Traditionally, wealthy people travelled from poorer countries, with basic health care facilities, to higher income countries that offered a better range of high-quality services. However, there has been a recent reversal in the direction of travel, with patients travelling from high-income countries in North America and Europe to low- and middle-income countries in Latin America and Asia (Horowitz and Rosensweig, 2007; Lautier, 2008). This new trend is driven by the ability of private facilities in lower income countries to offer high-quality services, with virtually no waiting time, at a comparatively low cost.

However, despite the 'hype' in the media regarding medical tourism, it is worth noting that the majority of this type of trade actually takes place regionally. For instance, patients from the US and Canada often go to destinations such as Brazil and Costa Rica; Western European patients travel to Eastern Europe; and patients from the Gulf countries and Pakistan travel to South and South East Asia, mainly India and Thailand. In addition to regional proximity, the country's specialty also plays a role in patients' decision of where to access care. This is because destination countries have specialised in certain procedures. For instance, Thailand and India specialise in orthopaedic and cardiac surgery, whereas Eastern European countries are hotspots for dental surgery (Smith et al., 2009).

There are significant limitations with the quantity and the quality of data available on medical tourism (Smith et al., 2009, 2011). Medical tourism from high- to low- and middle-income countries is currently an unregulated market that takes place through the private sector. Therefore, there is no systematic collection of data on how many patients travel, where they go to or what procedures they undergo. This is not to say that there is a lack of estimates in the literature, however. Data relating to numbers of medical tourists and profits that are made from this type of trade are often reported in newspaper and journal articles, and reports carried out by consultancy firms, such as McKinsey and Deloitte. However, the quality of these data is questionable. In terms of low- and middle-income countries the figures available in the literature suggest that medical tourism is happening on a large scale, and generating significant profits (Lautier, 2008; Smith et al., 2009).

This lack of reliable data poses problems for health care planners in both countries of origin and destination. Furthermore, it makes it very difficult to verify whether the proposed benefits and risks of this practice, which we review in the next section, actually occur. It is imperative that more empirical evidence is collected on medical tourism, including the numbers of patients and the effect this practice has on countries, particularly as there is currently much discussion on this topic, and it tends to be guided by ideology more than evidence (Smith et al., 2011).

Why do medical tourists travel to low- and middle-income countries?

The first step in exploring the phenomenon of medical tourism is to understand why people choose to travel to a country with lower levels of economic development than their own to access health services. The type of health system in the country of origin of medical tourists plays a key role in determining the reasons why patients travel. In countries with a market-driven health system, such as the US, the cost of health services is often the main reason for people to seek care abroad (Milstein and Smith, 2006; Burkett, 2007; Dunn, 2007). For instance, patients can save between 40 and 90 per cent by travelling from the US to India on procedures such knee surgery and heart bypass.[1] Patients who live in countries with a national, state-based, health service, such as predominate in Europe, often choose to travel to another country in order to bypass waiting times at their home health system, which can be months long for non-emergency procedures (Muzaffar and Hussain, 2007; Ramirez de Arellano, 2007; Turner, 2007b).

A further reason why people travel to low- and middle-income countries for health care is unavailability of certain procedures in their home country (Muzaffar and Hussain, 2007; Turner, 2007b; Connell, 2008). This often includes experimental treatments that have not been approved, such as stem cell therapy, or procedures that are deemed too expensive for the health system, such as hip resurfacing. Other reasons include increased confidentiality, for instance in cosmetic procedures (Horowitz et al., 2007), and for cultural similarities, where the Diaspora return 'home' for treatment, as they feel more comfortable with that health system (Reed, 2008).

Obviously, not all conditions lend themselves to medical tourism. It would not make sense, for instance, for someone with an emergency condition to board an eight-hour long flight to obtain care. Conditions that medical tourists travel for can therefore be classified as essential procedures, elective procedures and preventive care.

Essential procedures include cardiac surgery and transplants. Within cardiac surgery, many patients travel for cardiac bypass surgery, which countries like India and Thailand specialise in. There are disagreements over whether 'transplant tourism' should be treated as part of medical tourism, as the main driver of transplant tourism is the lack of organs available in many countries. However, transplant tourism raises a different set of ethical issues, as destination countries often have less regulated organ markets, and it is more difficult to ensure the well-being of the donor. Also, follow-up procedures may be more difficult, as the patient will need further treatment upon returning to their 'home' country (Canales et al., 2006; Turner, 2008).

By far the most popular procedures are elective ones. These include orthopaedic operations, which are often heavily priced in many high-income countries, and for which there are normally long waiting times, given that they are very common and typically non-emergency. Other types of elective procedures include eye surgery, dental treatment, gastric bypasses, cosmetic and gender reassignment surgery (Connell, 2006).

The final type of care patients travel for is preventive care. This represents a recent shift, as traditionally medical tourism was associated with curative care. However, healthy individuals are also travelling to have full body checkups, such as CT scans to ensure they do not have any 'hidden' conditions. These checkups are available in their countries of origin; however, they are often very expensive and not easy to access without defined clinical 'need'. This type of medical tourism has become popular in India, particularly when the Diaspora return 'home' for holiday or to visit relatives. There are issues associated with these types of service, as the radiation that patients expose themselves to during these scans is potentially harmful, and there may be an unnecessary level of distress for patients if they discover something that is relatively benign or that has no cure.

A final type of service medical tourists travel for are 'health farms'. These can act both as preventative care and recuperation services, and tend to be offered after surgery or full-body checkups to help recuperate or give 'health' advice such as eating and exercise habits to prevent disease. The number of these 'farms' has greatly grown in India over the past few years (Bhadoria and Mudaliar, 2009).

Medical tourism facilitators

Travelling to a low- or middle-income country for health care is far from being an easy and straightforward process. There are often language and cultural differences that may be hard to bridge, as well as lack of information about where to obtain treatment (both in terms of country and medical facility). This is especially the case in the current unregulated market, where most medical tourism takes place through patients arranging their own care, and paying out of their own pocket. This has given rise to medical tourism facilitators. There are different types of medical tourism facilitators, but they often act as the mediators between the hospital and the patient, and tend to be internet based. They can be divided into two types: those who provide only information to medical tourists and those who also arrange the care for them, often including flights and other forms of transportation, hotel accommodation, medical care and arrangements for an accompanying person. The need for these facilitators arises from a lack of information on the patients' side of what procedures are available and where. There have been many concerns regarding medical tourism facilitators, especially since they are not regulated (interestingly, travel agencies have stricter regulations!). Patients have no way of ensuring the information they are provided with is accurate, and there are concerns that there may be monetary incentives to portray hospitals in a positive light. Furthermore, medical tourism facilitators take no responsibility if the patients' experience is not satisfactory and have built in clauses that protect them against malpractice. There have therefore been recent calls in the literature for increased regulation of facilitators (Snyder et al., 2011).

How can low- and middle-income countries benefit from medical tourism?

Now that we have seen why patients travel abroad for health care, it is important to explore what countries stand to gain, or risk to lose, from engaging in medical tourism. While patients (and the countries they come from) can benefit from lower costs, faster access to care and increased confidentiality, low- and middle-income countries can also benefit from offering health services to foreign patients.

Revenues

As highlighted earlier, the most important reason why low- and middle-income countries engage in medical tourism is a financial one. Although, as indicated earlier, the data available on medical tourism are unreliable, the estimates of the benefits reported from some of the leading destination countries are very high (Timmermans, 2004; Ramirez de Arellano, 2007; Turner, 2007a). For instance, Thailand is reported to have generated US$1 billion in 2006 (Pachanee and Wibulpolprasert, 2006), and the medical tourism market in India was valued at over $310 million in 2005–06, but is predicted to generate US$2 billion by 2012 (Sen Gupta, 2008). Although medical tourism usually takes place through the private sector, the prospects of such high levels of foreign exchange have encouraged governments to invest in it directly or indirectly (through tax subsidies) (Lee, 2007; Ramirez de Arellano, 2007; Reed, 2008). For instance, the Indian government has considered medical tourism a 'deemed export', and has awarded it fiscal incentives, such as tax concessions, land at a subsidised rate and lower import duties (Sen Gupta, 2008).

As low- and middle-income countries often have over-stretched and under-resourced health sectors, this extra revenue can be invested back into the national health system and contribute to its improvement (Timmermans, 2004; Ramirez de Arellano, 2007). For instance, there are reports that in Thailand, India, Singapore and the Philippines medical tourism represents an important resource for economic and social development (Turner, 2007a). Moreover, this is often the subject of discussion by the Federation of Indian Chambers of Commerce and Industry (FICCI) and the Confederation of Indian Industries (CII) (personal communication with the authors). In addition, funds raised through medical tourism are not restricted to health care, but also include expenditure associated with the stay of accompanying persons or other tourism activities the patients may engage in while in the country.

'Brain drain' reversal

The second possible benefit from medical tourism is the opportunity to reverse their 'brain drain' (Chinai and Goswami, 2007; Dunn, 2007; Connell,

2008). The brain drain is a process by which health care professionals leave their countries of origin to work in other countries. This is often motivated by higher salaries and better career prospects. All types of health workers migrate, including doctors, nurses and pharmacists. The effects of this migration on the health system can be acute, as many low- and middle-income countries suffer significant staff shortages (Wibulpolprasert et al., 2004). Private hospitals offering health services to medical tourists are often able to provide health professionals with much better-paid jobs, and more promising career prospects, which has meant that some countries are therefore attracting health workers who migrated abroad to return home to work in these private hospitals (Chinai and Goswami, 2007). This has the additional benefit of giving a high-quality signal, as prospective medical tourists are more likely to trust doctors who trained or have practiced in their own country (Connell, 2008).

Quality of care

One of the key concerns of medical tourism is the quality of care patients receive (Martinez Alvarez et al., 2011; Smith et al., 2011). To address this, hospitals in exporting countries apply for international accreditation, ensuring the care they deliver is at least of the same quality of that offered in importing countries. Over 25 low- and middle-income countries have already obtained accreditation from the Joint Commission International, the international arm of the Joint Commission, which accredits US hospitals.[2] India, for instance, has had 19 hospitals accredited by them, while Thailand has accredited 28.

This raises standards of care in the hospitals, which also benefits the local population, as these hospitals do not exclusively treat foreign patients. Furthermore, hospitals catering for foreign patients are often required to offer a percentage of their beds for free to the local poor. Therefore, as this industry grows, the local population will also benefit from increased access to health care (however, as we see below, hospitals do not always observe these regulations).

Improved country image

Another advantage of medical tourism is improved country image. The more people that visit the country, and are happy with the services they receive, the better the country image will be, and the more general tourism and trade it will be able to attract.

What do low- and middle-income countries risk by engaging in medical tourism?

We now turn to the risks countries face when engaging in medical tourism. A literature review carried out on medical tourism identified the creation of a two-tiered health system as the chief concern for low- and middle-income countries (Smith et al., 2011). This would happen if foreign patients have access to sophisticated private hospitals that offer them state-of-the-art facilities and a high staff-to-patient ratio, when the local population only has access to basic facilities, often under-sourced, with low staff-to-patient ratios. This raises equity concerns, as it may be seen that the governments are 'subsidising' (or investing in) care for the wealthier foreign patients at the cost of the local population. Private hospitals often have a legal obligation to devote a percentage of their beds to the local poor for free. However, these regulations are rarely enforced (Sen Gupta, 2008).

Creation of an internal brain drain

The second concern medical tourism presents is the creation of an internal brain drain (Arunanondchai and Fink, 2006; Burkett, 2007; Chinai and Goswami, 2007). As we have already discussed, these countries can use medical tourism to attract health workers who have migrated to other countries. However, they risk creating an internal brain drain, where health professionals leave the public sector lured by the higher salaries and work opportunities provided by private hospitals catering for foreign patients. This can aggravate the problem of health worker shortage many of these countries are already experiencing, and again this raises concerns about equity.

Resource diversification

Medical tourists often have different health needs compared to the local population. Medical tourists require urban, tertiary care, whereas the majority of the local population needs rural primary health care. There is the danger that hospitals treating medical tourists, because they bring in higher revenues, may be prioritised over public health services, again leaving the local population worse-off (Chanda, 2002; Garud, 2005; Ramirez de Arellano, 2007; Connell, 2008; Leahy, 2008).

Utilisation of revenues

There is a worry that the local population will not benefit from the profits medical tourism brings into the country. This is because hospitals offering medical tourism services do not follow regulations that require them to devote a percentage of their beds to treating the poor free of charge. In addition, as in the majority of exporting countries, medical tourism occurs through the private sector, and only a very small proportion of the profits (if any) are spent on the public health system through taxes paid by these hospitals (Sen Gupta, 2008).

It is very important to note, however, that private hospitals in exporting countries do not cater for medical tourists exclusively. In fact, these only make up around 10 per cent of their patient base. The majority of patients in these private hospitals are the wealthier population of the exporting countries themselves. As such, medical tourism, rather than create the problems highlighted above, normally just exacerbates them, having a relatively marginal impact on an existing private sector (Smith et al., 2009).

Trade agreements

Under what legal conditions does this medical tourism occur? In this section we will explore the different mechanisms under which medical tourism takes place – multi-lateral, regional and bi-lateral – and how they can affect the impact medical tourism has on low- and middle-income countries.

The multi-lateral perspective

Patients organise their own travel privately within a multi-lateral framework, where countries trade with multiple countries under the same trade regulations. As a result, discussions on medical tourism are often framed within the World Trade Organisation's General Agreement on Trade in Services, or the GATS (Blouin et al., 2006). It is important to take into consideration that some of the issues associated with medical tourism only apply to this multi-lateral mode of trade. See the case study below for an example of the benefits and risks of medical tourism from the perspective of a different trade agreement. But let us first examine the GATS in a bit more detail, and what it means for health systems.

The GATS agreement was a result of the 1995 Uruguay Round Negotiations, and is the basis of the global multi-lateral sector trading system. GATS divides the service sector into four modes of trade. These are all applicable to the health service sector. Mode one covers cross-border supply of services, which in the case of health services refers to the remote provision of health services. Mode two is concerned with the consumption of services abroad, which refers to medical tourism. Mode three relates to commercial presence or foreign direct investment into another country's health system. Finally, mode four covers the presence of a natural person, which in the case of health systems refers to health worker migration. Under a GATS multi-lateral trade system, all countries trade with each other through one or more of these four modes. Often trade in health services also involves several modes simultaneously (Blouin et al., 2006; Smith et al., 2009).

Regional and bi-lateral trade

Although the debate concerning medical tourism tends to be framed within a multi-lateral context, this does not reflect the reality of the trade itself, where regional and bi-lateral trade agreements play a more prominent role (Smith et al., 2009). Regional travel would seem logical, as patients are more likely to travel shorter distances, and to places where there is a similar culture and little or no time difference. The Association of Southeast

Asian Nations, or ASEAN, region and the North American Free Trade Agreement, or NAFTA, in the Americas would be examples of regions where such travel takes place. For instance, a stakeholder analysis carried out in India revealed that the majority of the foreign patients that hospitals received came from neighbouring countries, particularly from the South Asian Association Regional Cooperation, such as Pakistan, Bangladesh and Afghanistan (Martinez Alvarez et al., 2011).

As far as bi-lateral trade is concerned there seems to be very little discussion taking place, both in the literature and in policy circles. This is concerning since, as we will see in the next section, bi-lateral trade has much to offer to countries wishing to engage in medical tourism (Martinez Alvarez et al., 2011).

A case study of a bi-lateral relationship between India and the UK

In this section we explore the issues we have discussed thus far, but from a bi-lateral trade perspective. We do this using India and the UK as a case study for this sort of trade, with a strong focus on how India, as the middle-income counterpart, would be affected by bi-lateral medical tourism. India and the UK are a good example for a bi-lateral trade relationship in medical tourism because they share a common language, a long history of intergovernmental relations and have similar health education systems.

India's health system is made up of a mix of public and private services. It has a low coverage, however, with only about 60 per cent of the population having access to health care (McKinsey and Confederation of Indian Industries, 2002). This results in a significant proportion of the population paying out of pocket, which has led to the development of a thriving private sector. In contrast, most of the health care in the UK is provided by the National Health Service, free of charge and available to all. Despite this, the UK National Health Service has come under increasing criticism due to its long waiting lists and large budgets, which has resulted in calls from the government for increased efficiency. In terms of trade in health services, the UK has already had some experience of this, primarily with other countries in the European Union. This includes having foreign doctors provide out-of-hours care (Freund et al., 2007), and patients being able to access care in other countries of the Union if they have not been offered suitable treatment within a reasonable amount of time by their home health system (Smith et al., 2011). Patients can also seek care outside of the European Union if the procedure they require is not available within the Union. Furthermore, there are estimates in the literature that up to a quarter of patients would be willing to travel abroad to access health care (Beecham, 2002).

Medical tourism in India

India is already a key player in other forms of trade in health services, such as the remote provision of health services and the movement of health care professionals. It is also an important destination of medical tourism. The Confederation of Indian Industries estimated that 150,000 Medical Tourists visited India in 2004 (McKinsey and Confederation of Indian Industries, 2002; Chinai and Goswami, 2007).

As we have already seen, the Indian government has granted medical tourism tax subsidies and land concessions, considering medical tourism a 'deemed export' in its 2002 National Health Policy (Garud, 2005; Burkett, 2007; Ramirez de Arellano, 2007; Sen Gupta, 2008). They have also introduced a medical visa, known as the 'M' visa, to facilitate the entrance of medical tourists to the country (Chinai and Goswami, 2007). Furthermore, Indian hospitals are very committed to obtaining national and international accreditation. The National Accreditation Board for Hospitals and Healthcare Providers has already accredited 117 hospitals.[3] As we saw earlier, 19 Indian hospitals have also already been accredited by the Joint Commission International.

There are many reasons for India's success in medical tourism. First, India has invested in world-class technical equipment, which together with the quality of health care professionals (many of whom have been trained abroad) equates to very high standards of care (Connell, 2006). Second, the price of care is considerably lower than in most high-income countries. Third, English is widely spoken, which gives it an advantage over countries such as Thailand, which needs to provide translation services at their hospitals (Connell, 2008).

Bi-Lateral relations between India and the UK

There is no formal bi-lateral trade in medical tourism taking place between the UK and India, although some papers have suggested the long waiting lists in the UK are a contributing factor towards India's success in medical tourism (Chatterjee, 2007; Sen Gupta, 2008). Some private players have started considering medical tourism as an option, such as the private health insurance company BUPA, who has an agreement with Ruby hospital in Kolkata, whereby British patients can be reimbursed for care. However, this is mainly for British nationals already residing in India (Connell, 2008). The Joint Economic and Trade Committee was a bi-lateral initiative set up in 2005 with the aim of removing trade barriers between India and the UK. It included a working group on health care, with some prospects of a medical tourism initiative, although this working group has been discontinued.

This lack of bi-lateral activity in health care should not be taken as a sign of lack of feasibility. A contract between the two countries could be drawn on medical tourism, and as we shall see has the potential of maximising the benefits and reducing the risks associated with this type of trade. In addition, there are several factors that would facilitate this relationship. First, there is a large Indian Diaspora population in the UK that would be willing to travel back 'home' for health care. Second, there are many doctors working in the UK's National Health System that are of Indian origin, and as such, the British population is already used to being treated by them. Last, the historical ties between these two countries can greatly facilitate this type of bi-lateral relationship.

Let us now explore how India could benefit from such a medical tourism relationship. A literature review exploring this case study found that provisions can be made on the contract signed by both countries to ensure that the profits made from this type of trade are incorporated into the public health system. In addition, the contract could pre-select hospitals that are following internationally acceptable standards, which could serve as an incentive for other centres to improve their services to reach the same standards (Smith et al., 2011).

Let us now turn to the risks posed by medical tourism, and how these can be minimised through a bi-lateral relationship. First, the contract could ensure better enforcement of regulations that allow the poor access to private facilities at no cost. If hospitals do not comply with these rules, they risk losing the contract with the UK National Health Service. In addition, a mechanism could be set up to ensure that the profits made from medical tourism benefit the local population. Again, mechanisms can be built into the contract signed by the two countries, whereby if these commitments are not adhered to, the contract will be stopped. Last, in order to tackle potential problems of the internal brain drain that may develop as a result of increased medical tourism, doctor exchange and training programmes could be established, either as part of the bi-lateral contract or as future programmes built on this relationship (Smith et al., 2011).

A study carried out by the authors interviewed stakeholders in both countries to assess the feasibility of this type of trade in medical tourism. This study showed that even though some stakeholders could see the benefits of such trade, the political cost was perceived to be too high (for instance, the government could not be seen to favour foreign patients at the expense of national ones). The concerns expressed differed between the two countries; for instance, stakeholders from the UK had concerns regarding the quality of care available in India, lack of continuity of care and regulation and litigation procedures. On the other hand, Indian stakeholders were mostly concerned about the impact this type of trade would have on the health care available to the local population, as well as how such a move would be perceived by the population. When a bi-lateral agreement was discussed, however, there was no consensus about whether and how the worries expressed would apply. This finding highlights the need for information on how much medical tourism is actually taking place and what impact it is having on countries providing these services and their health systems (Martinez Alvarez et al., 2011).

Case study conclusion

In conclusion, although medical tourism is often discussed from a multi-lateral, GATS-based perspective, a lot may be gained from countries engaging in a bi-lateral relationship. As we have seen in the example of India and the UK, a contract could be drawn between the two countries that pre-sets conditions so that benefits can be maximised and risks reduced.

Of course, although the UK and India were used as a case study here, this is by no means a unique example. There are many other countries that share similar historical and cultural ties that can engage in this type of agreement across the world, who are facing health systems challenges that can be addressed by engaging in bi-lateral medical tourism trade.

Conclusion

This chapter has considered why patients choose to travel to access health services in low- and middle-income countries and the impact this practice has on these countries. Although the data available are inaccurate, it is obvious from the literature that medical tourism to these countries is happening on an increasing scale. This can generate foreign exchange, reverse the brain drain and raise clinical standards, but risks exacerbating a two-tiered system and an internal brain drain. It is important to note that the different trade arrangements through which medical tourism can take place, namely multi-lateral, regional and bi-lateral, can have an influence on the impact medical tourism has, as shown by the case study on the potential for a bi-lateral relationship between the UK and India. Although trade agreements of this type are not currently taking place, there is reason to consider that they may be best placed to address the worries related to this practice. Finally, it is important to emphasise the need for more empirical data on the amount of medical tourism trade that is currently taking place and the effect it is having on all countries involved. At present, decisions are being made in a data vacuum, and are often more influenced by ideology than evidence.

Notes

1. http://www.medicaltourism.com/en/compare-costs.html
2. www.jointcommissioninternational.org
3. http://nabh.co/main/hospitals/accredited.asp

References

Arunanondchai, J. and Fink, C. (2006) 'Trade in health services in the ASEAN region' *Health Promotion International* 21, Suppl 1, 59–66.

Beecham, L. (2002) 'British patients willing to travel abroad for treatment' *British Medical Journal* 325, 10.
Bhadoria, B. M. S. and Mudaliar, A. (2009) 'SWOT Analysis On: Medical Tourism in India' *BSSS Journal of Management* 1, 70–83.
Blouin, C., Drager, N. and Smith, R. (2006) *International Trade in Health Services and the GATS: Current Issues and Debates* (The World Bank).
Burkett, L. (2007) 'Medical tourism. Concerns, benefits, and the American legal perspective' *Journal of Legal Medicine* 28, 223–45.
anales, M. T., Kasiske, B. L. and Rosenberg, M. E. (2006) 'Transplant tourism: Outcomes of United States residents who undergo kidney transplantation overseas' *Transplantation* 82, 1658–61.
Chanda, R. (2002) 'Trade in health services' *Bulletin of the World Health Organization* 80, 158–63.
Chatterjee, A. (2007) 'The art, science and commerce of medical tourism' *Journal of the Indian Medical Association* 105, 619–20.
Chinai, R. and Goswami, R. (2007) 'Medical visas mark growth of Indian medical tourism' *Bull World Health Organ* 85, 164–5.
Connell, J. (2006) 'Medical tourism: Sea, sun, sand and ... surgery' *Tourism Management* 27, 1093–100.
Connell, J. (2008) 'Tummy tucks and the Taj Mahal? Medical tourism and the globalization of health care' in A. G. Woodside and D. Martin (eds) *Tourism Management* (Oxon: CAB International).
Dunn, P. (2007) 'Medical tourism takes fight' *Hospital and Health Networks* 81, 40–2, 44.
Freund, T., Schwantes, U. and Lekutat, C. (2007) 'OOH care and locum doctors' *British Journal of General Practice* 57, 668–9.
Garud, A. D. (2005) 'Medical tourism and its impact on our healthcare' *National Medical Journal of India* 18, 318–9.
Horowitz, M. D. and Rosensweig, J. A. (2007) 'Medical tourism – health care in the global economy' *Physician Executive* 33, 24–6, 28–30.
Horowitz, M. D., Rosensweig, J. A. and Jones, C. A. (2007) 'Medical tourism: Globalization of the healthcare marketplace' *MedGenMed* 9, 33.
Lautier, M. (2008) 'Export of health services from developing countries: The case of Tunisia' *Social Science and Medicine* 67, 101–10.
Leahy, A. L. (2008) 'Medical tourism: The impact of travel to foreign countries for healthcare' *Surgeon* 6, 260–1.
Lee, C. (2007) 'Just what the doctor ordered: Medical tourism' *Monash Business Review* 3, 10–12.
Martinez Alvarez, M., Chanda, R. and Smith, R. D. (2011) 'The potential for bi-lateral agreements in medical tourism: A qualitative study of stakeholder perspectives from the UK and India' *Global Health* 7, 11.
Mckinsey and Confederation Of Indian Industries (2002) *Healthcare in India: The Road Ahead*, Confederation of Indian Industry (Ed.), no place of publication.
Milstein, A. and Smith, M. (2006) 'America's new refugees – seeking affordable surgery offshore' *New England Journal of Medicine* 355, 1637–40.
Muzaffar, F. and Hussain, I. (2007) 'Medical tourism: are we ready to take the challenge?' *Journal of Pakistan Association of Dermatologists* 17, 215–18.
Pachanee, C. A. and Wibulpolprasert, S. (2006) 'Incoherent policies on universal coverage of health insurance and promotion of international trade in health services in Thailand' *Health Policy Plan* 21, 310–8.

Ramirez De Arellano, A. B. (2007) 'Patients without borders: The emergence of medical tourism' *International Journal of Health Services* 37, 193–98.
Reed, C. M. (2008) 'Medical tourism' *Medical Clinics of North America* 92, 1433–46, xi.
Sen Gupta, A. (2008) 'Medical tourism in India: Winners and losers' *Indian Journal of Medical Ethics* 5, 4–5.
Smith, R. D., Chanda, R. and Tangcharoensathien, V. (2009) 'Trade in health-related services' *Lancet* 373, 593–601.
Smith, R., Martinez Alvarez, M. and Chanda, R. (2011) 'Medical tourism: A review of the literature and analysis of a role for bi-lateral trade' *Health Policy* 103, 276–82.
Snyder, J., Crooks, V. A., Adams, K., Kingsbury, P. and Johnston, R. (2011) 'The "patient's physician one-step removed": The evolving roles of medical tourism facilitators' *Journal of Medical Ethics* 37, 530–34.
Timmermans, K. (2004) 'Developing countries and trade in health services: Which way is forward?' *International Journal of Health Services* 34, 453–66.
Turner, L. (2007a) '"First world health care at third world prices": Globalization, bioethics and medical tourism' *BioSocieties* 2, 303–25.
Turner, L. (2007b) 'Medical tourism: Family medicine and international health-related travel' *Canadian Family Physician* 53, 1639–41, 1646–8.
Turner, L. (2008) '"Medical tourism" initiatives should exclude commercial organ transplantation' *Journal of the Royal Society of Medicine* 101, 391–4.
Wibulpolprasert, S., Pachanee, C. A., Pitayarangsarit, S. and Hempisut, P. (2004) 'International service trade and its implications for human resources for health: A case study of Thailand' *Human Resources for Health* 2, 10.

14

The Impact of the Internet on Medical Tourism

Daniel Horsfall, Neil Lunt, Hannah King, Johanna Hanefeld and Richard D. Smith

The key arguments of this chapter are as follows:

- Medical tourism has become increasingly prominent during the last decade and this is in large part a consequence of the way it is marketed through the Internet.
- Concerns regarding the quality of health information that exists on the Internet have been held for some time. Given that most medical tourism websites have a commercial purpose, assessing the quality of information available to consumers is extremely important.
- The findings of our study were as follows:

    - Commercial sites aimed at people seeking dental treatment abroad generally appear extremely professional.
    - The apparent professionalism of dental tourism sites masks the fact that important information is often missing from sites. Consequently dental tourism consumers are unlikely to be fully informed of all aspects of the dental tourism process.
    - The range of features adopted by dental tourism sites to engender trust are broad, though often meaningless.

- Addressing the poor quality of information on dental tourism sites is extremely important; however, for many reasons a regulatory approach is neither viable nor even desirable. It is likely that a combination of better education for patients as part of an ongoing relationship with 'offline' health care professionals, alongside self-regulation within the medical tourism industry represents the most realistic route forward.

Introduction

Within wider fields of patient mobility and health and wellness tourism one increasingly popular form of consumer movement is what has become commonly known as 'medical tourism' (Lunt and Carrera, 2010). This denotes what are typically non-emergency procedures driven by patients being more able – and willing – to travel to overseas destinations for specialised surgical treatments and other forms of medical care. With the rise of lower cost cross-border travel, rapid technological developments (encompassing both surgical techniques as well as the burgeoning volume and access to medical information and marketing on the Internet), these transactions are becoming more frequent and potentially serve a wide consumer market (Ehrbeck et al., 2008; Smith et al., 2009). However, although a key driver of this transnational travel, the role and quality of web-based information has not been thoroughly interrogated.

Medical tourism is the process where individuals opt to travel overseas with the *primary intention* of accessing medical treatments for which they intend to pay. These journeys include long-haul international travel, for example, from Western Europe to Asia, and encompass a range of treatments, including dentistry, cosmetic surgery, cardiac, bariatric and joint surgery, and fertility treatments (Heung et al., 2010; Johnston et al., 2010). A cursory glance at the Internet would suggest that the regions and countries at the forefront of medical tourism activities include Asia (Malaysia, Thailand and Singapore); Eastern Europe (Hungary and Poland); Mediterranean (Turkey and Cyprus); South and Central America (Costa Rica, Mexico, Brazil, and Cuba); and the Middle East (particularly Dubai and Jordon).

Arguably, there are two broad types of medical tourists, depending on the nature of the health system in question and how it is funded. First, medical tourists who may be categorised as 'consumers' because they use out-of-pocket and insurance payments to fund a range of dental, cosmetic and elective medical treatment (Carruth and Carruth, 2010; Smith-Morris and Manderson, 2010). Second, at the European level, medical tourism may also involve citizenship rights being invoked in order to receive medical surgery in another EU member state (Legido-Quigley et al., 2011; McHale, 2011; NHS, 2011).

The precise number of medical tourists making self-funded journeys is unknown, but the number of UK medical tourists has been estimated at about 100,000 per annum (Youngman, 2009). In terms of the global medical tourism market there are ongoing debates about the precise numbers seeking treatment, with estimates ranging from 80 thousand to five million annually (Ehrbeck et al., 2008; Youngman, 2009).

A key driver of medical tourism is the platform provided by the Internet for providing access to health care information, on the one hand, and advertising, on the other. The aim of this chapter is to provide a systematic review of medical tourism sites to help analyse their range and content, and begin to examine their quality and the implications of this in relation to decision making.

The chapter is structured into three sections and is organised as follows:

- An overview of medical tourism sites and the development of a conceptual framework to understand these;
- A detailed review of 50 commercial medical tourism websites for dental tourism using a checklist adapted from proposed questions consumers should ask according to a number of national-level bodies (NATHNAC, Department of Health; BDA). The final list was checked by an experienced dentist, familiar with litigation work;
- A consideration of the broader implications of these findings.

Overview of medical tourist websites

Bernstam et al. (2005) note that behind use for emails and product research, searching for health information is the third most common use of search engines. Similarly Lawrentschuck et al. (2012) highlight that as many as 86 per cent of all North-American internet users have at some stage browsed for health-related information. The precise nature of such internet usage is varied, ranging from searches for professional diagnosis to self-diagnosis, and from aftercare to support. Equally the precise nature of how people are using the Internet for health-related reasons varies considerably as does the format of the searcher–internet interface. Users might for example browse traditional web-pages simply for additional or background information, perhaps seeking a second or corroborating opinion of an 'offline' diagnosis. Some may use the Internet to purchase diagnostic services and treatments, while others may engage with or develop support networks online. This plurality of usage means that the tools experienced range from simple web-pages to more interactive discussion forums, file sharing sites and even closed community portals, each with a range of features ranging from the advertisement of products and services to, in some cases, the ability to chat 'live' with a doctor or expert.

The Internet is however not simply an increasingly interactive encyclopaedia offering information to surfers. Over the last decade the range of health treatments that have become available for purchase has dramatically increased and just as the Internet cannot be contained within national boundaries, nor can the products that are available to consumers. The Internet serves as a platform for gaining access to health care information and advertising and market destinations, and to connect consumers with an array of health care providers and brokers (Mainil et al., 2010). This means of course that important questions must be asked regarding the role and functioning of web-based resources, in particular the types and availability of information provided; the quality of this information; how patient confidentiality is respected and protected; and ultimately, how information provided over the Internet shapes patient choice of treatment, destination and provider.

There has been a burgeoning of sites dedicated to providing information for medical tourists in recent years. Although restricting its coverage to English-language websites, one scoping suggests a typology of websites can be drawn: (i) commercial portals (ii) media sites (iii) consumer-driven sites (iv) commerce-related sites and (v) professional contributions (Table 14.1, Lunt et al., 2010).

Medical tourism sites perform a range of functions, first and foremost the scope of the site is to introduce and promote services to the consumer. The main functionalities of the sites can be separated into five main processes: as a *gateway* to medical and surgical information, *connectivity* to related

Table 14.1 Types of medical tourism sites and their key features

Type of site	Features and notes
Portal	Provide an entry point to many destinations, brokers and providers. May cover one treatment type or many. Can be open, regionally focused, or focused on an individual provider. Use a range of features such as videos, testimonials and virtual tours. Ultimately provide consumer with opportunity to receive further information, often tailored to the individual and sometimes including quotes for treatments.
Media	Often these are focused on the marketing of the industry as a whole. Such sites can be presented as authoritative 'news' sites.
Consumer-generated	These include blogs and discussion boards. These often have input from both lay and professional interests. Such sites allow sharing of experience and sometimes a support community.
Commerce-related	Often peripheral to the medical tourism process, such sites offer information on commerce that is linked to the wider medical tourism industry. This can include cost-comparison sites, insurance sites, financial advice sites and even overseas property sales sites.
Professional, policy and regulatory-focused sites	These provide information on the regulation and legislation aspects of medical tourism. These are usually professionally developed and maintained and appear authoritative. Such authority is not always 'real' however.

health services, the *assessment* and/or promotion of services, *commerciality* and opportunity for *communication* (cf. Lunt et al., 2010).

The range of medical tourism sites and related content raise familiar concerns associated with unregulated online health information (Eysenbach, 2001; Scullard et al., 2010; Lichtenfeld, 2012). The sites are relatively cheap to set up and run, and contributors may post information without being subject to clear quality controls. A contextual deficit means selective information may be presented, or presented in a vacuum, ignoring, for example, issues such as post-operative care and support. There is also the possibility of unreliable products being marketed via the Internet – poor-quality surgery or inadvisable treatments, unnecessary and even dangerous treatments.

Clear evidence from other studies suggests that the quality of health information online is variable and should be used with caution. There have been some suggestions that the quality of information has been improving as a consequence of the increasing use of peer-reviewing of information on evermore interactive websites (Deshpande and Jadad, 2009); however it is still best characterised as variable (Langille et al., 2010; Honekamp and Ostermann, 2011; Lawrentschuck et al., 2012). In one study concerned with the quality of information pertaining to Inflammatory Bowel Disease on the Internet, researchers identified only 14 per cent of websites offering high-quality information (Bernard et al., 2007). Many of the websites they reviewed did not carry complete information, some carried incorrect information, key details regarding the source and currency of information were often lacking, and many sites required an advanced reading level to understand the content.

A review of patient-oriented Methotrexate information on the Internet found many of the same issues identified by Bernard and his colleagues, especially that the level of technical language was restrictively high (Thompson and Graydon, 2009). This study did however note that there were some examples of extremely high-quality information that was not only available but also returned on the first page of Google returns. One of the issues they identified was not that information supplied by websites was erroneous, rather some of the information experts would expect to be provided was simply missing. This issue is addressed by Scullard et al. (2010), whose research regarding the quality of online information pertaining to children's health revealed that only 39 per cent of 500 reviewed websites carried information to answer common health questions. Where an answer was offered, however, 78 per cent of all sites offered a complete and correct answer.

It is of course possible that missing information simply represents the specific focus of a website or perhaps even a reflection of frequently asked questions posed of a site. However there is also the scope to provide misinformation simply through the omission of important information. Mason and Wright (2011) suggest that this is a very real concern where medical tourism websites are concerned. Their review of 66 medical tourism sites noted a distinct over-selling of the positive aspects of medical tourism and a significant downplaying of risks.

Systematic review of commercial dental tourism sites

A sample of commercial websites offering treatment services abroad in the field of dental surgery was identified in the first instance. Building on published methodologies (Meric et al., 2002; Bernard et al., 2007; Thompson and Graydon, 2009; Scullard et al., 2010; Lawrentschuck et al., 2012) 50 sites were selected via the interrogation of the Google search engine with lay terms. While there exists variation in the precise methodology adopted, the literature is replete with examples of studies that adopted this broad approach. We believe that our samples represent an accurate reflection of the type of sites consumers are presented with when searching the Internet.

Drawing on previous studies we used (on 20 January 2012) the most popular search engine, Google (Laurent and Vickers, 2009; Scullard et al., 2010), to search for dental surgery offered abroad. Evidence suggests that interested searches are unlikely to travel beyond the first page of a search engine (Bernard et al., 2007; Laurent and Vickers, 2009; Scullard et al., 2010). As such, when using our primary search terms, we limited the number of returns to the first three pages, while additional search terms were limited to the first page of returns. We also followed the 'sponsored' advertisements that appear at the top of a Google search page as well as any advertisement banners that appear on the right-hand side of the search return page. In addition we followed a sample of links from any portal website that appeared in the first page of returns.

Our search of dental sites consisted of the major search terms: 'dental surgery', 'dental surgery abroad' and 'dentist abroad'. These returned a high number of relevant sites though there was much overlap. We supplemented these terms with further lay searches such as 'dental implant', 'dental implant abroad' , 'crowns abroad' and 'teeth whitening', as well as perhaps more technical searches such as 'veneers', 'cosmetic dentistry' and 'cosmetic dentistry prices'. This mirrored previous studies that focused on the quality of information on the Web pertaining to cosmetic surgery (Lunt and Carrera, 2010) and oncology information (Lawrentschuck et al., 2012).

All sites that were returned were included in our 50-site sample if they satisfied the criteria of being 'active', that is to say they had been updated recently, they offered surgery abroad directly through the website and they were not an extension of a site that had already been included. This means that we have used portals to access other websites but that the portals have not formed part of our sample. As a final check regarding the robustness of our sampling strategy the major search terms were entered into the search engines Bing, Yahoo and Alta Vista, where we found that the majority of our

Table 14.2 Destination locations promoted by the websites in the web review

Country location of dentist/clinic	Number of sites in sample
Hungary (mostly Budapest)	25
Ukraine	2
Turkey	2
Poland	3
Slovakia	1
Spain/Tenerife	2
Switzerland	1
Germany	1
Czech Republic	1

sites, including all those that appeared on page one in Google, were found in the first three pages of alternative search engines' returns.

Overall our sample consisted of four sites that were accessed through portals returned by the Google search, eight were not returned in the Google search but were followed through the banners on the right-hand side of the page and three sites were similarly accessed through adverts at the top of the Google return. The remaining 35 sites were listed in the Google returns, though there were a small number of these that also appeared either in a portal or as a banner. Of the 50 websites searched, only one did not work, only displaying a home page with broken links. For the 49 remaining websites, 4 were found to provide answers to 2 or less of the questions used to interrogate the sites.

In our sample a wide range of destinations were covered, though as Table 14.2 shows, Hungary was easily the most frequently advertised destination. The overall visual impact and content of sites varied widely. Most were professional, attractive and well-designed with alluring images of beauty, health and vitality; a small number however were extremely basic and did not reflect advanced web-design and marketing techniques and practices.

The major assessment involved the utilisation of a clinical checklist. To our websites we applied the following considerations from the perspective of a prospective medical tourist searching for information:

Clinicians

I. Is the dental practitioner named?
II. Are dental qualifications listed?
III. Is the dental practitioner a specialist?

Clinic

IV. Does the practitioner speak English/have translation?
V. Is the clinic equipped for medical emergencies?

VI. Are there any side effects or risks associated with the procedure?
VII. Who will be responsible for any clinical failures and their costs?

Regulation

VIII. Who is the practitioner's regulatory body?
IX. Does both the practitioner and clinic have professional indemnity cover?
X. Does the clinic have a complaints procedure?

Aftercare

XI. Will I be given a copy of my dental records on completion?
XII. What continuing care will be required?
XIII. Overall cost of the procedure.

This checklist was adapted from a series of similar checklists aimed at assessing the quality of information pertaining to other clinical areas such as cosmetic surgery (NaTHNaC, Department of Health; BDA) before being finalised with the help of a dental expert. We also supplemented this clinical checklist with a qualitative assessment of the non-clinical quality of the websites. Here the Health On the Net Code (HONCode) was extremely useful, providing us with a platform to assess how accessible, accurate and authoritative the websites are (see Meric et al., 2002; Khazaal et al., 2008; Lunt and Carrera, 2010 for examples of other studies that have made use of the HONCode).

Having developed the respective lists of criteria against which to judge our samples of sites, roles were divided within the five-person research team. Two members of the team performed the initial assessment of the dental websites against our checklist, while at the same time noting any interesting features of the websites. When this was completed the task was repeated by two more researchers who then came together as part of a moderation exercise with the initial team and any variations were explored. As a final check a fifth member of the research team then assessed a random sample of the sites and again, any disparity was discussed and addressed. Such procedures extend those adopted in similar methodologies (Thompson and Graydon, 2009; Scullard et al., 2010) in an attempt to avoid many of the limitations acknowledged by Meric et al. (2002).

Analysis of dental websites

Clinician Details: Out of the 50 websites reviewed 35 provide varying degrees of information about the clinicians employed within surgeries. Of these all named the dental practitioner and 31 gave details of the dentists' qualifications, often a full CV of qualifications and experience. However, the provision of information was not uniform – some simply stated a year and institution for qualification. In comparison, other websites provide extensive details on the dentist's credentials, including certification and current membership of professional bodies. Many websites (31 in total) enabled potential clients to ascertain whether the dentist is a specialist and if so, in what area.

Clinics: The majority of websites include photographs of the staff and the clinic, along with videos and even virtual tours of the facilities available. However a number of the photographs appeared somewhat random and unrelated to dentistry and many were duplicated either in different sections of the site or even on different websites (perhaps reflecting the interests of website builders). There were few sites giving exact details as to the number of dentists they employed, or the number of treatment rooms. For some, UK media coverage that they had received was seen as a major asset.

Language: Of our sample, 29 stated that their clinicians speak English and often other languages, most predominantly German and French. However, all of the websites accessed were written in English and there was an implicit suggestion that where information was presented in English, clients could assume that communication would not be a problem. The grammar and spelling were poor on a small number of sites, which is unlikely to reassure potential patients of the clinic or clinicians' ability to communicate proficiently in English.

When things go wrong: Very few websites acknowledge the possibility of something going wrong during treatment and it is therefore unsurprising that only 2 of the 50 websites explicitly said that their clinic was equipped for medical emergencies. Similarly, only five sites provide information about the possible side effects or risks associated with the different procedures that they carry out. It can be difficult for a prospective client to work out who would be held responsible if something does go wrong. Furthermore, only nine websites mention professional indemnity cover, but none provide explicit details, such as whether it covers both the practitioner and clinic.

Guarantees: None of the websites explicitly state that they will be responsible for clinical failures and any associated costs. However, 31 websites out of the 50 surveyed provide some form of guarantee for the dental treatment their clinic or dentist has provided. The guarantees vary for different types of treatment, but providers offer similar length of guarantee for comparable treatments across the sample. For example, dental implant guarantees range from 5 to 15 years, depending on the 'brand' of implant; bridges, crowns and veneers are normally guaranteed for around three years; and dentures and fillings for one year as standard. However, all guarantees come with certain conditions and exemption clauses. As standard, for the guarantees to be upheld, patients must return to the clinic (or the UK base/branch of the clinic) for a check-up, usually every 12 months, for an average fee of £50.

Most clinics offer to restore or repair dental work free of charge if there has been a problem that can be proven not to be the fault of the patient. However, all additional costs, including travel and accommodation, must be met by the patient. Box 14.1 outlines a number of other common circumstances in which guarantees will be invalidated, indicative of those included across the sample.

Box 14.1 Circumstances likely to invalidate guarantees

- neglected oral hygiene
- smoking or drug or alcohol abuse
- dentures, etc, are not maintained properly
- fail to attend the clinic for check-ups at least once every 6 months
- the clinic's instructions are not followed
- the gum tissue or teeth bone naturally recedes or reduces
- damage caused by stress
- substantial weight loss or gain within a short period of time
- contracting a general illness which negatively affects your mouth or teeth
- damage caused by accidents, for example, sports
- failure to pay the full cost of the treatment

There are also a number of procedures and circumstances that the guarantees do not cover. As standard, these include but are not limited to tooth-whitening procedures; pre-existing damage; root-canal treatment; and inappropriate selection of dental work by the patient.

Regulation: Of the sites 17 mentioned a regulatory body their practitioners are either registered with, or whose standards they adhere to. Interestingly even where a number of clinics operate in the same country, they cite different regulatory bodies and authorities. Sites frequently make statements pertaining to being regulated, but do not clarify the precise details. For example, one website states that it is a 'fully accredited British Dental Practice Abroad', while another claims to 'conform to and often exceed all the most stringent of EU regulations and requirements'. There were also 10 websites that declared they comply with ISO standards and two mention adhering to a commercial provider's Code of Practice.

Only 8 of the 50 sites mentioned a complaints policy and not all of these provide details of how a patient would go about complaining, or the course of action that would be followed.

Aftercare and follow-up: As a whole, websites provide very little detail about any aftercare that may be required following treatment. Twelve websites made some form of reference to aftercare, for eight of these it is simply to state that check-ups are available in their London clinic at extra cost. The remaining four make some basic suggestions, such as continuing dental hygiene.

Indeed, overall, sites sought to convey a sense of 'popping overseas' for a treatment holiday, something one site referred to as 'tweakending'. Only three of the sample websites mention patients' dental records and, of these, only two state that these are available on request. No details are offered on any of the other websites, potentially leaving potential clients wondering how future dental work would be informed about what they have already had done.

Treatment costs: The vast majority of sites did provide details about treatment costs (either Euros and/or Pounds Sterling). Only seven sites did not provide any information about costs. The cheapest service available was usually X-rays or tooth extraction, although some clinics provide X-rays for free and the most expensive treatments are consistently certain implants and bone reconstruction, usually from the hip. The price of treatments therefore ranged from as little as £2 to nearly £10,000 and sites often made reference to why their prices were so much cheaper than in the UK. Despite this, many of the websites advice patients to bring cash or traveller's cheques to pay for their treatment, implying that there may not be an official and traceable way of paying.

Discussion

Important questions persist around how consumers process the information they retrieve from website searches, how they take into account commercial interests and bias, and how this all contributes towards decision making. While our study does not tackle these dimensions there is some evidence that dental patients or prospective consumers of dental services do indeed use the Internet for a range of purposes. In their study Ni Riordain and McCreary (2009) report that over a third of respondents ($n = 500$) had sought information regarding a presenting oral or dental problem while similar proportions suggested they would consult an online dental practitioner and potentially use the Internet to plan trips abroad for dental treatment. A similar study by Harris and Chestnutt (2005) had found much lower proportions (three per cent) of respondents to have accessed dental information and only two per cent of respondents had purchased dental products (including treatment) through the Internet. They did however report that 44 per cent of respondents indicated that they would be interested in such opportunities, especially if directed to such sites by their dentist, which led them to conclude that there was real merit to expanding the provision of dental health information.

With regard to the UK it is clear that dentists and their regulatory bodies are becoming increasingly aware of the role of information and advertising on the Internet, which is reflected in the establishment of the GDC's (2012) 'Principles of Ethical Advertising' guideline, which has recently been enforced in the face of some criticism. Elsewhere, Peterson et al. (2003) suggest consumers of medicine are aware of issues around bias, commercialisation and lack of regulation when they explore health sites, but suggest that the context of what is being searched is important. They argue that commercial considerations 'may have an impact on the motives for and quality of information'. What is unclear, for example, is whether potential consumers purposely seek information that cautions about possible pitfalls and difficulties, as well as some of the more aesthetic and clinical attractions of medical tourism (Birch et al., 2007). It is important to know far more about how those who are accessing medical tourist information judge the information they retrieve given such information may be confusing, overwhelming, and even contradictory.

The marketing of medical tourist treatment raises considerations and concerns of direct-to-consumer advertising. Gollust et al. (2003) examine the direct-to-consumer internet sales of genetic services and note that sites are likely to exaggerate benefits – a finding mirrored by Mason and Wright (2011) who looked at a wide range of medical tourism sites. Illes et al. (2004) focus on direct-to-consumer advertising in print and information brochures, concluding that such materials fail to provide consumers with the sort of comprehensive and balanced information necessary for informed decision making. They suggest it is common to identify misinformation, unsubstantiated scientific claims, fear-provoking threats and a lack of information on the uncertainties and the risks of particular services (in their case tomographic and magnetic resonance imaging). They suggest that on the whole websites do not mention long-term risks or problems, limiting themselves to acknowledging that further treatment may be necessary as a consequence of hitherto unknown circumstance. A typical statement that illuminates this last point being one that suggests that if 'it turns out you have bad roots, you may need root canal treatment as well'.

In line with these findings, commercial sites in our sample seek to raise awareness of medical tourism, create a perceived need and ensure the consumer is motivated to purchase. There is, however, a major imbalance between the types of site identified with most information available from sites with clear commercial imperatives. In particular reviews of web content have shown that many dental websites carry unclear or out-of-date information, false claims regarding the specialisms held by dentists and in some cases offers to provide 'Botox' treatment and 'teeth whitening' despite legal restrictions on such advertising (Addy et al., 2005; Nichols and Hassall, 2011). The stance taken by the GDC reflects the fact that within the UK, health care for example has not traditionally been viewed as simply another product to be marketed and advertised.

There exists then a general appreciation of the existing and potential problems related to health information on the Internet, especially given the increasing quantity of health-related websites. Issues of accuracy are longstanding, though perhaps compounded by the burgeoning number of sites whose primary purpose is not providing information, rather selling products. What is less clear is how (if at all) the quality of medical tourism information is best addressed with suggestions ranging from codes of conduct; self-taken quality labels; user guidance tool; third-party quality and accreditation labels (Deshpande and Jadad, 2009); to educating users and assisting those wishing to search (Scullard et al., 2010; Lawrentschuck et al., 2012; Lichtenfeld, 2012).

The most common approach to assessing the standard of information is to employ some form of quality measurement checklist as we have done here. This has in many cases led to suggestions that formalising such tools and better educating prospective consumers to their value represents the most realistic approach to ensuring that patients are able to assess the quality of information they are being subjected to. However, which forms of accreditation or badges of quality control should be adopted will always provoke debate (See Bernstam et al., 2005; Deshpande and Jadad, 2009; and Thompson and Graydon, 2009 for some examples). Most agree that the HON code is a good start, but the need to add clinical-area-specific questions means that consumers are going to spend a considerable amount of time deciding which quality standard to look for before they have even begun their search for health information or products. And as Bernstam et al. (2005) note, few of the near three hundred quality-rating instruments that existed seven years ago were likely to be practically usable to lay audiences.

One study has suggested that a fusion of an expert system and a meta-search engine might enable consumers to access high-quality information that has been verified by experts, through a web-based portal (Honekamp and Ostermann, 2010). While their pilot study suggested that this enabled lay participants to access high-quality yet understandable information, two problems remain. First, the quality is guaranteed by the meta-search restricting to those websites that are 'accredited', in this case via the HoN label. Again, this would require an unambiguous agreement regarding which quality-rating instrument was best suited. Secondly and on a more practical level, HoN refuse to certificate medical tourism websites.

Most studies also recommend that there is a better education of prospective e-patients or e-health consumers by their 'offline' medical professionals. Lawrentschuck et al. (2012) go as far as to suggest that professionals should take the time to provide a steer to ethical, accurate, readable and accessible sites. This is echoed by Scullard et al. (2010) and both studies make the recommendation that if health care professionals can direct patients to government websites or other such reputable sources of information, patients will be able to minimise risk. There are of course a few problems with this approach, the first being that many prospective patients become that on the basis of what they find on the internet, rather than having visited a healthcare professional offline. Second, even when potential e-patients/consumers have visited healthcare professionals, it is not necessarily the case that these professionals will be aware of reputable sites beyond perhaps the broad and often general sites such as NHS Direct. And third, some of the sites one might assume conform to quality-based expectations do indeed fall short (Addy et al., 2005; Nichols and Hassall, 2011).

One suggestion is to simply move on from the debate about the quality of health information on the Internet. Deshpande and Jadad (2009) conclude in much the same fashion as Bernstam et al. (2005) – that the latest iteration of a quality-rating instrument (in this case the Medication Website Assessment Tool) shares many of the limitations of other such tools. They suggest that much debate that surrounds the quality of information reflects a now outdated understanding of the Internet. The new features of many websites brought about by the advent of Web 2.0 have, they insist, inspired a 'health 2.0' to emerge. They argue that the increased interactivity through the use of blogs, wikis and podcasts, often provided by not just health care professionals, but also researchers and the general public, has helped formulate a new e-health era that is tailored to the needs of those who access it. Crucially though this 'bottom-up' approach involves a process that is analogous to peer-reviewing, drawing on real-life experiences, but always ensuring that an expert source is nearby. Deshpande and Jadad cite the example of 'Wikipedia', which despite relying on anonymous, unpaid volunteers is, they say, as accurate in its coverage of scientific information as the 'Encyclopaedia Britannica' (2009, p. 2).

Deshpande and Jadad are convinced that the increasingly interactive nature of the Internet will create an informed patient or consumer community and it is to be hoped that they are correct. However, the findings of our study, alongside others mentioned in this chapter suggest that at least for the uninitiated, those new to searching the Internet for health information, there is a lot of poor-quality information that is likely to be found, and discerning what is useful and what is not is no easy task.

Conclusion

Whether online or offline individuals are faced with decisions about their health on a regular basis. In many ways such decisions reflect the increasing element of choice available to people and undoubtedly reflects the rise of a more consumerist, post-Fordist society. But whereas the choices regarding which goods or services should be consumed is generally a positive experience to the benefit of the consumer, we cannot ignore the fact that health is not a commodity in the same sense as other products. While most consumers will undertake some form of research when considering a purchase, many such decisions are of little consequence. Of course the greater the risk of a loss (financial or otherwise) the greater the level of research an individual is going to undertake and in turn the higher likelihood that third-party information will be sought, processed and utilised. Where one's health is at stake, it comes as no surprise that individuals seek recourse to health professionals and other sources of medical information.

The increasing attention paid to and availability of medical tourism is yet another example of the expanding level of choice 'enjoyed' by consumers. But it also underscores the fact that individuals are having to take yet more responsibility for their decisions, decisions that have potentially much more serious consequences than most other consumer choices. The rise of medical tourism has been a largely online phenomenon and it is the Internet that provides the arena in which all the key stakeholders, from the patients, through the facilitators, to the providers, are connecting. But this is not necessarily a safe arena; in the offline world where such decisions are taken in doctors' surgeries and hospitals there is at least a sense that information is authoritative, impartial and trustworthy. The Internet does not offer this and it is hard not to see the marketing interests behind even the most reassuring information found on the Internet. With regard to the websites we reviewed there were many affirmations of integrity, quality and caring services. There was an abundance of guarantees of professionalism, safety and a continuation of care. Yet there can be no guarantee that such promises are not empty from the very beginning, or at least become so as soon as the patient is discharged. How we reconcile the potential of the Internet, its endowment of individuals with ever-increasing levels of choice, with the risks inherent with an unregulated and uncontrollable content in a way that empowers patients is unclear and undoubtedly not straightforward.

It is perhaps easy and misleading to think of the relationship between patients and health care providers having fundamentally and irrevocably changed as a consequence of the Internet. Even now there is a feeling that word-of-mouth is still the main driver behind new patients registering with dental practices, and this relationship of trust between professional and patient may hold the key to how the Internet can be best harnessed for patients, with trusted professionals playing a larger role in providing health care information online. What is clear is that the goal of tempering the flow of information on the Internet is unrealistic. As it seems developing a system of accreditation or quality-rating though a benchmark/criteria may be an option that various stakeholders may be amenable to.

References

Addy, L. D., Uberoi, J., Dubal, R. K. and McAndrew, R. (2005) 'Does Your Practice Website Need Updating?' *British Dental Journal* 198, 259–60.

Bernard, A., Langille, M., Hughes, S., Rose, C., Leddin, D. and Veldhuyzen van Zanten, S. (2007) 'A Systematic Review of Patient Inflammatory Bowel Disease Information Resources on the World Wide Web' *The American Journal of Gastroenterology* 102, 2070–7.

Bernstam, E. V., Shelton, D. M., Walji, M. and Meric Bernstam, F. (2005) 'Instruments to Assess the Quality of Health Information on the World Wide Web: What Can Our Patients Actually Use?' *International Journal of Medical Informatics* 74, 13–19.

Birch, J., Caulfield, R. and Ramakrishnan, V. (2007) 'The Complications of 'Cosmetic Tourism': An Avoidable Burden on the NHS' *Journal of Plastic, Reconstructive and Aesthetic Surgery* 60, 1075–7.

Carruth, P. J. and Carruth, A. K. (2010) 'The Financial and Cost Accounting Implications of Medical Tourism', *International Business and Economics Research Journal* 9, 8, 135–40.

Deshpande, A. and Jadad, A. (2009) 'Trying to Measure the Quality of Health Information on the Internet: Is it Time to Move On?' *Journal of Rheumatology* 36, 1–3.

Ehrbeck, T., Guevara, C. and Mango, P. D. (2008) Mapping the Market for Medical Travel. *The McKinsey Quarterly* [Online], https://www.mckinseyquarterly.com/Mapping_the_market_for_travel_2134 date accessed 3 April 2012.

Eysenbach, G. (2001) 'What is E-health?' *Journal of Medical Internet Research,* 3, e20.

GDC (2012) *Principles of Ethical Advertising* (London: GDC).

Gollust, S. E., Wilfond, B. S. and Hull, S. C. (2003) 'Direct-to-consumer Sales of Genetic Services on the Internet' *Genetics in Medicine* 5, 332–7.

Harris, C. E. and Chestnutt, I. G. (2005) 'The Use of the Internet to Access Oral Health-related Information by Patients Attending Dental Hygiene Clinics' *International Journal of Dental Hygiene* 3, 70–3.

Heung, V. C. S., Kucukusta, D. and Song, H. (2010) 'A Conceptual Model of Medical Tourism: Implications for Future Research' *Journal of Travel and Tourism Marketing* 27, 236–51.

Honekamp, W. and Ostermann, H. (2010) 'Anamneses-Based Internet Information Supply: Can a Combination of an Expert System and Meta-Search Engine Help Consumers find the Health Information they Require?' *Open Medical Informatics Journal* 4, 12–20.

Honekamp, W. & Ostermann, H. (2011) 'Evaluation of a prototype health information system using the FITT framework'. *Informatics in Primary Care,* 19, 47-49.

Illes, J., Kann, D., Karetsky, K., Letourneau, P., Raffin, T. A., Schraedley-Desmond, P. Koenig, B. A. and Atlas, S. W. (2004) 'Advertising, Patient Decision Making, and Self-Referral for Computed Tomographic and Magnetic Resonance Imaging' *Archives of Internal Medicine* 164, 2415–9.

Johnston, R., Crooks, V. A., Snyder, J. and Kingsbury, P. (2010) 'What Is Known About the Effects of Medical Tourism in Destination and Departure Countries? A Scoping Review' *International Journal for Equity in Health* 9, 24.

Khazaal, Y., Fernandez, S., Cochand, S., Reboh, I. and Zullino, D. (2008) 'Quality of Web-Based Information on Social Phobia: A Cross-Sectional Study' *Depression and Anxiety* 25, 461–5.

Langille, M., Bernard, A., Rodgers, C., Hughes, S., Leddin, D. and van Zanten, S. V. (2010) 'Systematic Review of the Quality of Patient Information on the Internet Regarding Inflammatory Bowel Disease Treatments' *Clinical Gastroenterology and Hepatology* 8, 322–8.

Laurent, M. R. and Vickers, T. J. (2009) 'Seeking Health Information Online: Does Wikipedia Matter?' *Journal of the American Medical Informatics Association* 16, 471–9.

Lawrentschuck, N., Sasges, D., Tasevski, R., Abouassaly, R., Scott, A. and Davis, I. (2012) 'Oncology Health Information Quality on the Internet: a Multilingual Evaluation' *Annals of Surgical Oncology* 19, 706–13.

Legido-Quigley, H., Passarani, I., Knai, C., Busse, R., Palm, W., Wismar, M. and McKee, M. (2011) 'Cross-border healthcare in the European Union: clarifying patients' rights' *British Medical Journal,* 342, 364-367.

Lichtenfeld, L. J. (2012) 'Can the Beast Be Tamed? The Woeful Tale of Accurate Health Information on the Internet' *Annals of Surgical Oncology* 19, 701–2.

Lunt, N. and Carrera, P. (2010) 'Medical Tourism: Assessing the Evidence on Treatment Abroad' *Maturitas* 66, 27–32.

Lunt, N., Hardey, M. and Mannion, R. (2010) 'Nip, Tuck and Click: Medical Tourism and the Emergence of Web-Based Health Information' *The Open Medical Informatics Journal* 4, 1–11.

Mainil, T., Platenkamp, V. and Meulemans, H. (2010) 'Diving into the Contexts of In-Between Worlds: Worldmaking in Medical Tourism' *Tourism Analysis* 15, 743–54.

Mason, A. and Wright, K. B. (2011) 'Framing Medical Tourism: An Examination of Appeal, Risk, Convalescence, Accreditation, and Interactivity in Medical Tourism Web Sites' *Journal of Health Communication* 16, 163–77.

McHale, J. V. (2011) 'The New EU Healthcare Rights Directive: Greater Uniformity?' *British Journal of Nursing* 20, 442–4.

Meric, F., Bernstam, E. V., Mirza, N. Q., Hunt, K. K., Ames, F. C., Ross, M. I., Kuerer, H. M., Pollock, R. E., Musen, M. A. and Singletary, S. E. (2002) 'Breast Cancer on the World Wide Web: Cross Sectional Survey of Quality of Information and Popularity of Websites' *British Medical Journal* 324, 577–81.

NHS (2011) 'Patient Choice Beyond Borders: Implications of the EU Directive on Cross-Border Healthcare for NHS Commissioners and Providers' *Briefing* (Brussels: NHS European Office).

Ni Riordain, R. and McCreary, C. (2009) 'Dental Patients' Use of the Internet' *British Dental Journal* 207, 583–6.

Nichols, L. C. and Hassall, D. (2011) 'Quality and Content of Dental Practice Websites' *British Dental Journal* 210, E11–E11.

Peterson, G., Aslani, P. and Williams, K. A. (2003) 'How do Consumers Search for and Appraise Information on Medicines on the Internet? A Qualitative Study Using Focus Groups' *Journal of Medical Internet Research* 5, e33.

Scullard, P., Peacock, C. and Davies, P. (2010) 'Googling children's health: reliability of medical advice on the internet' *Archives of Disease in Childhood* 95, 580–2.

Smith, R. D., Rupa, C. and Viroj, T. (2009) 'Trade In Health-Related Services' *The Lancet* 373, 593–601.

Smith-Morris, C. and Manderson, L. (2010) 'The Baggage of Health Travelers' *Medical Anthropology: Cross-Cultural Studies in Health and Illness* 29, 331–5.

Thompson, A. E. and Graydon, S. L. (2009) 'Patient-Oriented Methotrexate Information Sites on the Internet: A Review of Completeness, Accuracy, Format, Reliability, Credibility, and Readability' *The Journal of Rheumatology* 36, 41–9.

Youngman, I. (2009) 'Medical Tourism Statistics: Why Mckinsey Has Got It Wrong' *International Medical Travel Journal*, http://www.imtj.com/articles/2009/mckinsey-wrong-medical-travel/ date accessed 3 April 2012.

15

Towards a Model of Sustainable Health Destination Management Based on Health Regions

Tomas Mainil, Keith Dinnie, David Botterill, Vincent Platenkamp, Francis van Loon and Herman Meulemans

This chapter:

- Introduces the idea of a destination management framework for transnational health care.
- Considers the definitions and concepts that inform an analysis of transnational health care, governance and sustainability.
- Presents the building blocks of destination management, specifically stakeholder, ethical and branding theories.
- Demonstrates how the linkages between destination management and transnational health care are constructed.
- Demonstrates how regional development in relation to health and health care is an active practice in the EU.

Transnational health care, cross-border health care, governance and sustainability

Health care is subject to the processes of both globalisation and commodification (Relman, 1980; Pelligrino, 1999). Related to these phenomena is the movement of international patients across regions and around the globe. In the European Union (EU) a large part of this patient mobility is being regulated by the European Directive on Patients' Rights (Council of the European Union, 2011). It is from this perspective that the term cross-border health care (CBHC) should be considered. In Europe there is a regional approach towards patient mobility (Brand et al., 2008). Border regions have been and are examples of good practice where cross-border health care, the provision side of patient mobility, takes place within the EU. Cross-border health care has always been pursued from a patient-driven perspective (Glinos et al., 2010). It also incorporates other forms of patient mobility, such as tourist accidents that require immediate treatment and senior citizens residing in southern destinations (Legido-Quigley et al., 2012) who need health care.

In contrast to the EU, in other parts of the world commodification has made a more decisive impact on the daily practices of patient mobility. From an Anglophone (Snyder et al., 2011) and Asian (Ormond, 2011) perspective, the tendency is to refer to medical tourism (industry-based) and medical travel (scholarly accepted). These terms originate from an industry which is based on a different, much more supply driven business model. Or as Glinos et al. (2010) explain, 'rather than focusing on the suppliers of health care and their interests in patient mobility... the industry-driven term "medical tourism" insinuates leisurely travelling and does not capture the seriousness of most patient mobility'.

The term 'transnational health care' (Mainil et al., forthcoming) originates from the notion that transnational organisations in health care such as private and public insurance schemes are working across borders and in national health systems. Just like large corporate hospital chains, insurance schemes are also active globally. This transnationalism could be professionalised further and networks could be developed further, but in close coordination with public health authorities or governments. Such transnational activities could create a more visible presence within health care that could better serve the interests of the international patient, with proper surveillance by public health authorities. Therefore we see transnational health care as a future, developed, globalised practice where integration could take place between EU examples of cross-border health care and Anglophone medical tourism practices, bringing together industry and public health perspectives.

With regard to setting the institutional frameworks and legal governance in relation to patient mobility in the EU, the EU Directive on Patients' Rights (Council of the European Union, 2011) provides patients with more opportunities for mobility, for health and economic reasons, but at the same time offers and reserves an important role for the member states in guiding patient mobility. This includes the instalment of reference centres, measures of quality and information infrastructure. In the light of these developments one could also ask the following question: 'Is there room for regional steering?' Could regional governments, governing a health region where a large part of the workforce is active in the health care and well-being market, where there is a motivation to attain an identity as being a health region, where there are enough stakeholders present to steer this identity and where there is a joint concern for the local and transnational community, obtain more authority to decide which centres of excellence are being installed in their region, so that these initiatives are made accountable as part of a whole health region? Regions could then choose which specific medical (and well-being) areas to focus on, with a specific set of providers and facilities, based on the needs of (trans)national populations and the wants of public health authorities, creating regions with a specific set of medical (and well-being) expertise. Other nearby regions could create another set of expertise. This means that health region development becomes a driver for the management of the medical supply side in the EU taking into account the needs of the patient and resident community. Such a strategy is not without its critics, as indicated by previous practices in countries such as the UK and its National Health Service (NHS). Its impact on travel time for both patients and their families within such a specialised model of health-care provision and the perceived loss of local services affected during its development phase are both difficult, and unresolved, political issues. Similar developments can be found in the logic and effects of mergers of hospitals (Fulop et al., 2002; Gaynor et al., 2012).

However, using this model, steering power is given to regional governments by a (supra-)national authority and regional development for patient mobility becomes a sustainable element of public health provision. We propose that both national and international patient groups will benefit from these health regions, because the regions can provide a better pool of medical expertise and organisational efficiency and thus be promoted more adequately, both nationally and internationally. This regional approach is in line with more transnational corporate organisations (Dicken, 2009), who could partake in the regional activities as stakeholders. Sustainability is therefore approached here as a consequence of governmental intervention. If the measures benefit the populations as patients present in the EU, then it is a sustainable practice, as opposed to a more commodified approach, subject to privatisation and based on profit. Sustainability as such is seen here in a way similar to the way 'sustainable tourism' (Gössling et al., 2008; van der Duim and Caalders, 2008) has been conceived: tourism activity with a view of the local community. Sustainable health destination management (SHDM) should serve the public health goals and targets of EU (supra-)national governments.

So far, we have explored several concepts important for patient mobility and have introduced a regional approach for this phenomenon. We will now explain 'destination management' as an instrument for patient mobility governance and show how some EU regions are developing practices that refer to this instrument. Finally, we will relate the current debates on EU patient mobility to our proposed framework.

Destination management and transnational health care

Destination management may be conceptualised as a practice operating at four geographic levels: city, domestic region, country and international

Table 15.1 Levels of destination management

Destination level	**Examples**
City	Helsinki, Barcelona (Hospers, 2011)
Domestic region	Spanish (17) and Italian regions (19) (Barrutia and Echebarria, 2010)
Country	Australia (Baker, 2011)
International region	English-Welsh cross-border region (Ilbery and Saxena, 2011)

region (Table 15.1). For example, Hospers (2011) has described the city-level strategies undertaken by Helsinki and Barcelona; Barrutia and Echebarria (2010) have studied domestic region-level efforts in Spain and Italy to build their social capital, research and development, and innovation capacities; Baker (2011) has shown how at nation level the destination management organisation (DMO) of Australia has sought to integrate cultural expression in the form of film into the branding of the country; and at international region level, Ilbery and Saxena (2011) have analysed the strategic, administrative and personal challenges inherent in managing integrated rural tourism approaches in the English-Welsh cross-border region.

Although the destination management literature has focused primarily on leisure-based tourism (Mistilis and Daniele, 2004; Pike, 2004; Dwyer et al., 2009), in recent years there has been increasing interest in the field of medical tourism (Connell, 2005; Bies and Zacharia, 2007; Turner, 2007; Cohen, 2008; Hazarika, 2010). With the steady rise in the number of individuals travelling abroad for medical services (Keckley, 2008), over 50 countries now promote themselves as international destinations for medical services (Woodman, 2008). As Mainil et al. (2010, p. 753) have stated for a part of globalised populations, '...the notion of local health care provision is slowly being left behind. A new global form of health care perception is on the rise and the shift has already occurred in the tourism field.' This has led to an increased interest in patient mobility from both policy makers and citizens. It therefore seems timely to examine the theoretical foundations of destination management in order to gain insights into how and to what extent destination management techniques may be applicable – or are actually already in use (Ormond, forthcoming) – to the domain of transnational health care. The theoretical foundations of destination management may be viewed as threefold, comprising perspectives drawn from stakeholder theory, ethical theory and branding theory. This holistic view of destination management (Dinnie, 2008) underlines the importance of treating the subject as a complex social phenomenon rather than merely as a managerial practice (Figure 15.1).

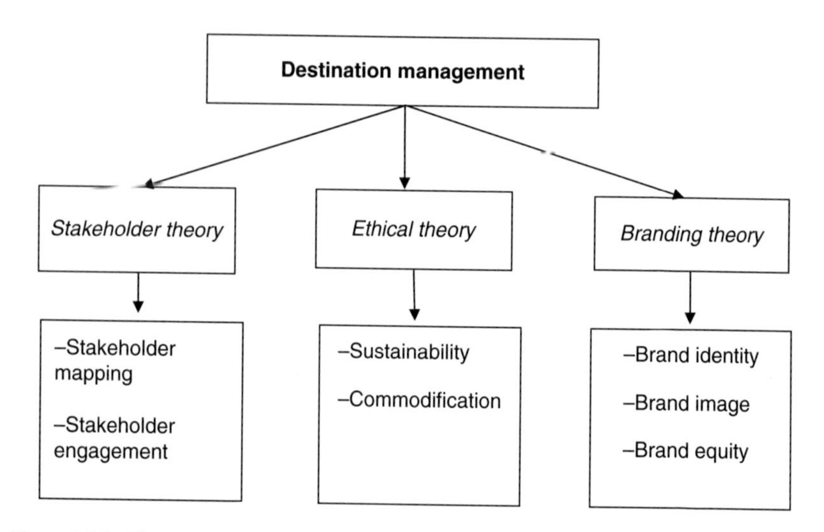

Figure 15.1 Theoretical foundations of destination management

Stakeholder theory

Destination management entails the participation of multiple stakeholders in the pursuit of achieving shared goals. In the context of corporate organisations, stakeholders have been defined as 'groups that are themselves affected by the operations of the organisation, but can equally affect the organisation, its operations and performance' (Cornelissen, 2004, p. 59). One can question, following the framework of patient mobility, whether citizens/patients would be included in this definition of stakeholders. Within the general management literature, several studies have attested to the importance for organisations to develop and maintain a stakeholder orientation (Greenley and Foxall, 1997; Christensen, 2002; van Woerkum and Aarts, 2008).

Given the perceived importance of stakeholders, organisations of all kinds need to identify and then engage with their various stakeholder groups and individuals. Mitchell et al. (1997) have suggested that organisations may grade the salience of their stakeholders in terms of the power, legitimacy and urgency of each group. In an examination of Mitchell et al.'s power, legitimacy and urgency framework, Parent and Deephouse (2007) found that power has the most important effect on salience, followed by urgency and legitimacy. In the context of sustainable tourism development, it has been argued that 'in order to produce equitable and environmentally sustainable tourism developments multiple stakeholders must be involved in the process of planning and implementing the project' (Currie et al., 2009, p. 41). This indicates the scope of the challenge inherent in adopting an inclusive approach to stakeholders – multiple stakeholders not only need to be included in the initial planning phase of a project but also need to be engaged in the project's implementation. Policy makers need to ensure that adequate resources are allocated to manage such a wide-ranging, inclusive process.

Cross-border health care and medical tourism are moving towards transnational health care as a future professional field, involving more professional structures and networks, for example providing information to prospective transnational patients and creating transnational follow-up care structures, which would benefit from greater attention using a stakeholder approach. Governments could play an important role in organising a set of stakeholders to address patient mobility from a holistic, meaning inclusive/integrative, perspective. In setting the region as a framework within which different stakeholders interact, medical (hospital providers) and non-medical stakeholders (insurance providers and intermediaries) could benefit. As an example, we have observed that in Malaysia the government plays an important strategic role in positioning the country and its domestic regions with regard to patient mobility, alongside the role of the private sector in this endeavour (Ormond, forthcoming).

Ethical theory

Ethical theory contributes to our understanding of destination management by drawing attention to two key concerns – the need to address the issue of 'commodification' and the importance of ensuring that the development of the destination is conducted in a sustainable manner. The question of commodification revolves around concerns over how appropriate it is to 'package' and promote a destination using techniques similar to those employed by companies for their products. Arguing beyond commodification leads to the distribution of resources to different social classes in society. Health as a resource or social good is also commodified (Pelligrino, 1999). The risk is that commodification will reduce the perceived authenticity and value of a place as a destination (Klieger, 1990; Dearden and Harron, 1992). However, this view has been challenged by other researchers who contend that commodification of a place can play a positive role in the local community. For example, Abram (1996, p. 198) suggests viewing commodification as 'part of a very positive process by which people are beginning to re-evaluate their history and shake off the shame of peasantry'. A more contemporary view is expressed by Cole (2007, p. 945–6), who suggests that 'understood from the perspective of the local people, cultural commodification can be positive... it needs also to be recognised as part of a process of empowerment'. Commodification is not without controversy in many arenas (including health care), it can lead not only to empowerment but also to disempowerment of citizens within the spectrum of the development. Particular sensitivity needs to be displayed towards the issue of commodification when the focus of attention is a phenomenon as integral to well-being as is transnational health care.

Developing a destination in a sustainable manner is key to the advancement of successful transnational health care. Tourism researchers have established a perspective according to which '...the success of area-based sustainable tourism development...largely lies in providing local actors with the means to engage in a process of managing complex and potentially risky situations; the collective learning of stakeholders, supporting multiple-loop learning, is a key mechanism for arriving at more desirable (sustainable) futures' (Koutsouris, 2009, p. 577). The increasing awareness of environmental issues has led to sustainability becoming an important component of destination management (Chang, 2009). Building on this recognition of the centrality of sustainability in destination management, Insch (2009) has conceptualised green destination brands in terms of the manner in which destinations emphasise the green dimensions of their brand identity. It should be made clear that sustainability in the context of this paper and regarding patient mobility is not so much concerned with acting 'green', but more to do with good population public health, activated by several levels of governmental policy.

Commodification is present in transnational health care. This means that destination management could counterbalance the excesses of commodification by installing control by governments on the development of global health-care initiatives and ensuring an equitable relationship with the public health-care systems serving local populations. Transnational health care would benefit from sustainable solutions with regard to the resident populations and communities. As an example, Health care Belgium[1] is representing a framework for patient and professional mobility for Belgium, but needs to take into account the Belgian government and possible reactions of the public opinion. The general issue, if enhanced patient mobility and the flow of more foreign patients into the Belgian health system is wished for, is still under debate.

Branding theory

The increasing level of competition between destinations to attract medical tourists has heightened the importance of DMOs' attempts to differentiate themselves from the competition through the development of distinctive place branding strategies. As an academic discipline, place branding has relatively recent roots (Gertner, 2011). The application of branding techniques to places is grounded upon three key concepts in the branding literature – brand identity, brand image, and brand equity.

The first of these concepts, brand identity, has been defined as 'a set of aspects that convey what a brand stands for: its background, its principles, its purpose and ambitions' (van Gelder, 2003, p. 35). Brand image, on the other hand, refers to the mental perceptions that people hold of the brand, regardless of how accurately those perceptions reflect the reality of the brand. In the field of tourism, destination image has been the focus of considerable research (Pike, 2002). Image is considered to be central to tourists' destination evaluation and selection process (Echtner and Ritchie, 1993). The third of these concepts, brand equity, is well established in the marketing and branding literature and frequently transferred to the tourism literature. It is treated from a consumer perspective as customer-based brand equity, defined as 'the differential effect that brand knowledge has on consumer response to the marketing of that brand' (Keller et al., 2012, p. 54). The customer-based brand equity concept has been applied to destinations by Konecnik and Gartner (2007), who emphasise that in addition to image, other important dimensions include awareness, quality and loyalty.

Regions and governments, engaged in transnational health care, would be able to create awareness as health regions, by branding them as sustainable health regions, where a balance is present between commodification and the public good of the population. If regional governments get the chance to choose the global health-care initiatives they visualise for their regional development, this would mean they could also promote their region as a healthy one for (trans)national populations, with a well-balanced supply, based on access to health care, of medical and well-being providers. This would provide better integration between nearby regions, installing regional development as a tool for enhanced patient mobility and further developing transnational health care. As an example, Bavaria, in Germany, is promoting itself as 'Bavaria, a better state of health'[2], combining health and environmental policy, and promoting itself as a health region: a combination of medical and well-being facilities.

Case-studies: Health care Belgium, Maas-Rhine region, the Veneto region and Bavaria

Each of the cases is examined against the triad of stakeholder, ethical and branding theories, identified above as inherent to destination management. This will indicate the differences in character of the regions and assess which regions are closer to the concept of sustainable health destination management. These regions were selected on the basis of their involvement in patient mobility developments.

Health care Belgium, Belgium

Stakeholder perspective: Health care Belgium is a private non-profit organisation, sponsored by stakeholders (such as university and private hospitals and medical suppliers), but in combination with influential governance required by the Belgian government. On the basis of these characteristics we can say that they are adopting a stakeholder approach: working together with governments/health insurance providers/employers and hospitals. The organisation is engaged in delivering secondary and tertiary health care on the basis of bilateral contracts with other regions and nations. This involves outsourcing Belgian medical expertise to these regions/nations and taking care of specialised patient cases in Belgium. A similar recent development could be observed when Belgium offered aid to those wounded in the war in Libya. However, at the present time stakeholders are not fully committed to a joint governance model; they are still very much competitors, inherent to the traditions and practices of the distribution of resources within the medical sector.

Ethical/local community perspective: The Belgian citizens' cultural experience of their health system makes patient mobility still a difficult exercise. Among local populations there exists fear regarding the quality and capacity of Belgian hospitals. Because Belgium is a country that follows the continental European social model, it takes a persistent political standpoint which means that opening the borders to foreign patients is sensitive, and that a two-speed health-care system must at all times be resisted in favour of the Belgian population.

Branding/promotional perspective: From an international health-care perspective, Belgium is somewhat known, but it has not yet profiled itself particularly effectively. Health care, being a social good and right, has never been seen as a potential industry. A difference can be observed in reference to the case study of Bavaria below. However, in terms of a 'health region', it is argued that the area surrounding Brussels would have the opportunity to be a health region, because of the extent of medical expertise, easily accessible airports, the presence of many embassies, quality hotels, and having the overall international character of a capital city. In conclusion, Belgium, and Brussels in particular, could develop into a sustainable health region if policy levels were structurally more open for managing patient mobility.

Maas-Rhine region, the Netherlands

In this region there are several stakeholders investing in solving health issues but the regional government is not driving this dynamic. The consequence of this is that they are in less of a position to steer the region towards becoming a health region. However, there are other powerful stakeholder groups who are influential. Using a top-down perspective, the EUREGIO (a cross-border cooperation organisation) Maas-Rhine organisation operates at several policy levels in order to influence public health matters. On the other hand, there are bottom-up organisations such as EUPrevent which include several stakeholders, but from a public health prevention perspective. Additionally, hospitals such as the university hospitals of Maastricht (AZM) and Aachen (UK Aachen) are trying to establish centres of excellence but have difficulty in promoting themselves in other policy levels.

Ethical/local community perspective: One can observe a gap between the reality of cross-border patient mobility in the Maas-Rhine region and the inherent conservatism within the political establishment, which at the moment is more against than for the EU.

Branding/promotional perspective: This region has been seen for years as a region where cross-border patient mobility plays a significant role. This is known to several stakeholders in Europe, but it is less seen as such by the resident population. The political context also plays a role here. The Dutch government is not in favour of large volumes of mobile patients and other EU initiatives, due to the present political constellation. In conclusion, this is a region which is developing slower than previously, but with organisations such as EUREGIO and EUPrevent acting as steering bodies there remains an opportunity to develop further as a health region.

Veneto region, Italy

Stakeholder perspective: In the 1990s 20 regions were created. Ninety per cent of hospitals are controlled by these regions and regional policies are strongly linked to the local economy. The region is an important tourist destination which places demands on local services, including health services. For example, this year there were 20,000 usages of a cross-border insurance card for minor/major emergencies. There is, therefore, already close alignment between extant regional economic and health policy. Close connections with bordering regions are conceived as important for this regional government: in particular with Austria, Slovenia and Croatia. The present health-care system means that there are no gains from increasing the volume of international patients because the majority of hospitals are publicly funded. However, the spectre of the EU Patients' Rights Directive will mean adapting to new challenges. This means new chances to prepare medical staff for international patient care and to obtain quality accreditation. Concerning the role of stakeholders, if a hospital is interested in attracting international patients, then the regional government needs to sign bi-lateral agreements, to prove commitment, and to offer a framework for local administrators. In this way, the involvement of regional government is assured. It is observed that in regions with a strong dependency on tourism the case for the involvement of a wider set of stakeholders is more easily made.

Ethical/local community perspective: The preoccupation of Veneto citizens are linked to length of waiting times for treatment in emergency rooms, given the large amounts of tourists arriving every year in the Veneto region. The government is responding by investing in primary care provision and health promotion.

Branding/promotional perspective: The promotion of the hospitals of Veneto is lacking at this stage. There is little awareness, on a pan-European basis, of the quality of health care in Veneto; consequently Veneto is known as a tourism destination rather than as a health region as such. A recent tendency is that hospitals that opened 2–3 years ago are beginning to promote themselves as medical treatment centres but these actions are taken independently. In conclusion, the Veneto regional government is committed to health care but the focus of its response to tourism is very much related to the demands of tourists for emergency care. However, the region's image as a well-known tourist destination would be a good starting point from which to develop as a transnational health region.

Bavaria: 'a better state of health', Germany

Stakeholder perspective: A lot of the efforts of the Bavarian state department of health and economic development are centred around the enhancement of the medical chain. Greater efficiency is needed to effect improvements for the local as well as for international patients. They also have signed bilateral contracts with other nations to enhance patient mobility. They are actively developing new integration initiatives. For example, they have provided funds to and accredited with a quality mark 17 (of 70 counties) health regions in Bavaria, based on their commitment to providing health care. These are regions which depend on a public health economy, as a large part of their economic value lies with the health-care sector. The state government funds them to create networks and cooperation between several stakeholders. As such, the government has enabled the players to work together and to accommodate organisational and management functions, literally under one roof. There is strong cooperation with local political officials who are charged to make the cluster networking work.

Ethical/local community perspective: The importance of the public health economy is being exemplified because a significant part of the population works in this sector. At the same time, the state government needed to search for cost efficiency for the medical sector in Bavaria. One of the policies adopted was to encourage hospitals to work closer together by clustering services in order to improve quality. It also involved the development of medical specialties at specific hospitals.

Branding/promotional perspective: With the strong strategy 'Bavaria, a better state of health' and an underlying logic of the public health economy, it seems that making connections between different stakeholders and being present in the right spaces has been beneficial for the region of Bavaria to position itself as a sustainable health region. In conclusion, the importance of the public health economy and its treatment as an industry and the involvement of state government in developing the region as a health region connects Bavaria to the concept of sustainable health destination management.

Taking a comparative overview of the four case study regions, several differences and stages of development are apparent. Belgium and the

Maas-Rhine region share the condition that governmental support is less forthcoming in the current political context, although these regions do have characteristics which are conducive to sustainable health destination development. The Veneto region is more dependent on the regional government, although it is driven more by its notoriety as a tourist region, but with health system implications. Finally, the Bavarian case shows how integration and cooperation of stakeholders, steered by specific initiatives and solid commitment from the government, can produce a health region approach. The type of governance demonstrated in Bavaria is very near the model of SHDM and transnational health region development that we are proposing. Although the regions are different from each other, they share a joint characteristic – of having a certain openness towards mobility and a willingness to be more than a local region. Practices such as the EUREGIO in the Maas-Rhine region, Health Care Belgium, the installation of health regions in Bavaria, and the large health-care infrastructures in the Veneto region, show an ability to anticipate the changes occurring across borders.

Discussion: Health regions and cross-border health care in Europe

As a further contribution to the implications of applying destination management to patient mobility frameworks, we have analysed how this is related to the current EU debate on cross-border health care. In this section we examine how our ideas about health region development sit against EU health policy considerations (see also Chapter 9).

Firstly, following Glinos et al.'s (2010) observations on cross-border health care in Europe: what is the willingness of EU populations to consume health services abroad? Secondly, there is a factor of significant socio-economic differences in Europe, which has an influence on patient mobility and the related national policies. Finally, the current economic climate could also not be the most beneficial for the development of economical patient flows (with a tendency towards localised versus globalised patients). Developing transnational health regions could enhance patient mobility and the willingness of EU populations to travel for their health, but it could also constrain the willingness to travel because of the socio-economic differences among EU nations and regions (for example Hungarian citizens not being able to travel for health reasons to Germany). The development of a health region is also an economic activity which could be beneficial if it leads to cost-efficiency, but overall not all EU national governments are in the position to invest in such an initiative.

When arguing in terms of the EU Directive on Patients' Rights, which is the latest approach of the EU commission to govern patient mobility in Europe, a set of scenarios can be stipulated around the potential of the model of health region development (Mainil et al., forthcoming).

Scenario 1: The current regulations under the EU Directive will not create the dynamics for more cross-border mobility. This means that health region development will only be accidental and based on individual governmental decisions. The framework of the Directive is not seen as an opportunity to build capacity by governments for their citizens.

Scenario 2: Current regulations (EU Directive on Patients' Rights) are implemented: this means that national governments have to argue and act in terms of management of European and/or international patient flows which opens the possibility to think in terms of health region development, with the government as a mediator, as is occurring now in the Bavarian case.

Scenario 3: Implementation creates a dynamic on several stakeholders – such as insurance providers/hospitals/supportive services/governmental structures/citizens – to become transnational, resulting in transnational health regions. In this case it is not enough if only the regional and national governments steer the health region development. Supra-national bodies such as the European Commission also need to steer the development of health regions. This involves taking decisions on which medical chain initiatives (a steered process of several providers in the delivery of medical services) should be embedded in which health regions. For the three scenarios different levels of capacity building can be employed in developing the supra-(national) public health systems. This could include transnational health region development that incorporates an increased capacity within public health provision that serves both local and transnational patients.

Conclusion

This chapter has sought to establish both a conceptual and a practice-based link between the ideas implicit in the terms 'destination management' and 'transnational health care'. This takes the form of a model of SHDM that potentially results in transnational health region development. Several regional cases have been explored within the framework of the factors of destination management (stakeholder–ethical–branding theory). Finally, we have considered our proposals within the current challenges facing cross-border patient mobility in the EU. Future research needs to include how the EU Directive on Patients' Rights is implemented and how its effects could be assessed. Furthermore, an assessment of whether regions in Europe are or tend to be transnational health regions, and how they relates to public health provision, would be an interesting line of research.

Acknowledgments

We would like to express our gratitude to the professionals for their insights on the specific regions.

Notes

1. http://www.healthcarebelgium.com/
2. http://www.state-of-health.bayern.de/

References

Abram, S. (1996) 'Reactions to Tourism: A View from the Deep Green Heart of France' in J. Boissevain (ed.) *Coping with Tourists. European Reactions to Mass Tourism* (Oxford: Berghahn Books).

Baker, B. (2011) 'Branding and the Opportunities of Movies: Australia' in N. Morgan, A. Pritchard and R. Pride (eds) *Destination Brands – Managing Place Reputation* (Oxford: Butterworth-Heinemann).

Barrutia, J. M. and Echebarria, C. (2010) 'Social capital, research and development, and innovation: An empirical analysis of Spanish and Italian regions' *European Urban and Regional Studies* 17, 4, 371–85.

Bies, W. and Zacharia, L. (2007) 'Medical tourism: Outsourcing surgery' *Journal of Mathematical and Computer Modelling* 46, 1144–1159.

Brand, H., Hollederer, A., Wolf, U. and Brand, A. (2008) 'Cross-border health activities in the Euregios: Good practice for better health' *Health Policy* 86, 245–54.

Chang, T-Z. (2009) 'Sustainability as Differentiation Tool in Destination Branding: An Empirical Study' *3rd International Conference on Destination Branding and Marketing* (Institute for Tourism Studies, Macao SAR, China, 2–4 December 2009) 121–29.

Christensen, L. T. (2002) 'Corporate communication: The challenge of transparency' *Corporate Communications: An International Journal* 7, 3, 162–8.

Cohen, E. (2008) 'Medical tourism in Thailand' *AU-GSB e-journal* 1, 1, 24–37.

Cole, S. (2007) 'Beyond authenticity and commodification' *Annals of Tourism Research*, 34, 4, 943–60.

Connell, J. (2005) 'Medical tourism: Sea, sun, sand and . . . surgery' *Tourism Management* 27, 1093–100.

Cornelissen, J. (2004) *Corporate Communications – Theory and Practice*. (London: Sage Publications).

Council of the European Union. *Directive on Cross-border Health Care Adopted*. Brussels, 7056/11, PRESSE 40; 2011.

Currie, R. R., Seaton, S. and Wesley, F. (2009) 'Determining stakeholders for feasibility analysis' *Annals of Tourism Research* 36, 1, 41–63.

Dearden, P. and Harron, S. (1992) 'Case Study: Tourism and the Hill Tribes of Thailand' in B. Weiller and M. Hall (eds) *Special Interest Tourism* (London: Belhaven).

Dicken, P. (2009) 'Globalization and transnational corporations' *International Encyclopedia of Human Geography* 563–69 (Oxford: Elsevier).

Dinnie, K. (2008) *Nation Branding – Concepts, Issues, Practice* (Oxford: Butterworth-Heinemann).

van der Duim, V. R. and Caalders, J. D. A. D. (2008) 'Tourism chains and pro-poor tourism development: An actor-network analysis of a pilot project in Costa Rica' *Current Issues in Tourism* 11, 2, 109–25.

Echtner, C. and Ritchie, B. (1993) 'The measurement of destination image: An empirical assessment' *Journal of Travel Research* 31, 4, 3–13.

Dwyer, L., Edwards, D., Mistilis, N., Roman, C. and Scott, N. (2009) 'Destination and enterprise management for a tourism future' *Tourism Management* 30, 1, 63–74.

Fulop, N., Protopsaltis, G., Hutchings, A., King, A., Allen, P., Normand, C. and Walters, R. (2002) 'Process and impact of mergers of NHS trusts: Multicentre case study and management cost analysis' *British Medical Journal* 325, 7358, 246.

Gaynor, M., Laudicella, M. and Propper, C. (2012) 'Can governments do it better? Merger mania and hospital outcomes in the English NHS' *Journal of Health Economics* 31, 3, 528–43.

Gertner, D. (2011) 'A (tentative) meta-analysis of the "place marketing" and "place branding" literature' *Journal of Brand Management* 19, 2, 112–31.

Glinos, I. A., Baeten, R., Helbe, M. and Maarse, H. (2010) 'A typology of cross-border patient mobility' *Health and Place* 16, 6, 1145–55.

Gössling, S., Hall, C. M. and Weaver, D. (eds) (2008) *Sustainable Tourism Futures Perspectives on Systems, Restructuring and Innovations* (London: Routledge).

Greenley, G. E. and Foxall, G. R. (1997) 'Multiple stakeholder orientation in UK companies and the implications for company performance' *Journal of Management Studies* 34, 2, 259–84.

Hazarika, I. (2010) 'Medical tourism: Its potential impact on the health workforce and health systems in India' *Health Policy and Planning* 25, 248–51.

Hospers, G-J. (2011) 'City branding and the tourist gaze' in K. Dinnie (ed.) *City Branding –Theory and Cases* (Basingstoke: Palgrave Macmillan).

Ilbery, B. and Saxena, G. (2011) 'Integrated rural tourism in the English-Welsh cross-border region: An analysis of strategic, administrative and personal challenges' *Regional Studies* 45, 8, 1139–55.

Insch, A. (2009) 'Green Essence or Green Wash? Conceptualisation and Anatomy of Green Destination Brands' *3rd International Conference on Destination Branding and Marketing* (Institute for Tourism Studies, Macao SAR, China, 2–4 December 2009) 190–8.

Keckley, P. H. (2008) *Medical Tourism: Updates and Implications* (Washington, DC: Deloitte Center for Health Solutions).

Keller, K. L., Apéria, T.and Georgson, M. (2012) *Strategic Brand Management: A European Perspective* 2nd edn (Harlow, England: FT Prentice Hall).

Klieger, P. (1990) 'Close Encounters: "Intimate" Tourism in Tibet' *Cultural Survival Quarterly* 14, 2–5.

Konecnik, M. and Gartner, W. C. (2007) 'Customer-based brand equity for a destination' *Annals of Tourism Research* 34, 2, 400–21.

Koutsouris, A. (2009) 'Social learning and sustainable tourism development; local quality conventions in tourism: A Greek case study' *Journal of Sustainable Tourism* 17, 5, 567–81.

Legido-Quigley, H., Nolte, E., Green, J., la Parra, D. and McKee, M. (2012) 'The health care experiences of British pensioners migrating to Spain: A qualitative study' *Health Policy* 105, 1, 46–54.

Mainil, T., Platenkamp, V. and Meulemans, H. (2010) 'Diving into the context of in-between worlds: Worldmaking in medical tourism' *Tourism Analysis* 15, 6, 743–54.

Mainil, T., Van Loon, F., Dinnie, K., Botterill, D., Platenkamp, V. and Meulemans, H. (In Press) 'Transnational health care: Towards a global terminology' *Health Policy* 108, 1, 37–44.

Mistilis, N. and Daniele, R. (2004) 'Challenges for competitive strategy in public and private sector partnerships in electronic national tourist destination marketing systems' *Journal of Travel and Tourism Marketing* 4, 63–73.

Mitchell, R. K., Agle, B. R. and Wood, D. J. (1997) 'Toward a theory of stakeholder identification and salience: Defining the principle of who and what really counts' *Academy of Management Review* 22, 4, 853–86.

Ormond, M. (2011) 'Shifting subjects of health care: Placing "medical tourism" in the context of Malaysian domestic health-care reform' *Asia Pacific Viewpoint* 52, 247–59.

Ormond, M. (2013) *Neoliberal Governance and International Medical Travel in Malaysia* (London: Routledge).

Parent, M. M. and Deephouse, D. L. (2007) 'A case study of stakeholder identification and prioritization by managers' *Journal of Business Ethics* 75, 1, 1–23.

Pelligrino, E. D. (1999) 'The commodification of medical and health care: The moral consequences of a paradigm shift from a professional to a market ethic' *Journal of Medicine and Philosophy* 24, 3, 243–66.

Pike, S. (2002) 'Destination image analysis: A review of 142 Papers from 1973 to 2000' *Tourism Management* 23, 5, 541–9.

Pike, S. (2004) *Destination Marketing Organisations: Bridging Theory and Practice (Advances in Tourism Research)* (Oxford: Elsevier Science).

Relman, A. S. (1980) 'The new medical-industrial complex' The *New England Journal of Medicine* 303, 17, 963–70.

Snyder, J., Crooks, V. A., Adams, K., Kingsbury, P. and Johnston, R. (2011) 'The "patient's physician one-step removed": The evolving roles of medical tourism facilitators' *Global Medical Ethics,* 8 April 2011.

Turner, L. (2007) 'Medical tourism: Family medicine and international health related travel' *Canadian Family Physician* 53, 1639–41.

van Gelder, S. (2003) *Global Brand Strategy: Unlocking Brand Potential Across Countries, Cultures and Markets* (London, UK: Kogan Page).

van Woerkum, C. and Aarts, N. (2008) 'Staying connected: The communication between organizations and their environment' *Corporate Communications: An International Journal* 13, 2, 197–211.

Woodman, J. (2008) *Patients Beyond Borders: Everybody's Guide to Affordable, World-class Medical Tourism* (Chapel Hill, NC: Healthy Travel Media).

Index

Printed and bound in the United States of America